London Mathematical Society
Lecture Note Series 63

Continuous Semigroups in Banach Algebras

ALLAN M. SINCLAIR

CAMBRIDGE UNIVERSITY PRESS

LONDON MATHEMATICAL SOCIETY LECTURE NOTE SERIES

Managing Editor: Professor I.M.James,
Mathematical Institute, 24-29 St Giles, Oxford

1. General cohomology theory and K-theory, P.HILTON
4. Algebraic topology: a student's guide, J.F.ADAMS
5. Commutative algebra, J.T.KNIGHT
8. Integration and harmonic analysis on compact groups, R.E.EDWARDS
9. Elliptic functions and elliptic curves, P.DU VAL
10. Numerical ranges II, F.F.BONSALL & J.DUNCAN
11. New developments in topology, G.SEGAL (ed.)
12. Symposium on complex analysis, Canterbury, 1973, J.CLUNIE & W.K.HAYMAN (eds.)
13. Combinatorics: Proceedings of the British combinatorial conference 1973, T.P.McDONOUGH & V.C.MAVRON (eds.)
14. Analytic theory of abelian varieties, H.P.F.SWINNERTON-DYER
15. An introduction to topological groups, P.J.HIGGINS
16. Topics in finite groups, T.M.GAGEN
17. Differentiable germs and catastrophes, Th.BROCKER & L.LANDER
18. A geometric approach to homology theory, S.BUONCRISTIANO, C.P.ROURKE & B.J.SANDERSON
20. Sheaf theory, B.R.TENNISON
21. Automatic continuity of linear operators, A.M.SINCLAIR
23. Parallelisms of complete designs, P.J.CAMERON
24. The topology of Stiefel manifolds, I.M.JAMES
25. Lie groups and compact groups, J.F.PRICE
26. Transformation groups: Proceedings of the conference in the University of Newcastle upon Tyne, August 1976, C.KOSNIOWSKI
27. Skew field constructions, P.M.COHN
28. Brownian motion, Hardy spaces and bounded mean oscillation, K.E.PETERSEN
29. Pontryagin duality and the structure of locally compact abelian groups, S.A.MORRIS
30. Interaction models, N.L.BIGGS
31. Continuous crossed products and type III von Neumann algebras, A.VAN DAELE
32. Uniform algebras and Jensen measures, T.W.GAMELIN
33. Permutation groups and combinatorial structures, N.L.BIGGS & A.T.WHITE
34. Representation theory of Lie groups, M.F.ATIYAH et al.
35. Trace ideals and their applications, B.SIMON
36. Homological group theory, C.T.C.WALL (ed.)
37. Partially ordered rings and semi-algebraic geometry, G.W.BRUMFIEL
38. Surveys in combinatorics, B.BOLLOBAS (ed.)
39. Affine sets and affine groups, D.G.NORTHCOTT
40. Introduction to H_p spaces, P.J.KOOSIS
41. Theory and applications of Hopf bifurcation, B.D.HASSARD, N.D.KAZARINOFF & Y-H.WAN
42. Topics in the theory of group presentations, D.L.JOHNSON
43. Graphs, codes and designs, P.J.CAMERON & J.H.VAN LINT
44. Z/2-homotopy theory, M.C.CRABB
45. Recursion theory: its generalisations and applications, F.R.DRAKE & S.S.WAINER (eds.)
46. p-adic analysis: a short course on recent work, N.KOBLITZ
47. Coding the Universe, A. BELLER, R. JENSEN & P. WELCH
48. Low-dimensional topology, R. BROWN & T.L. THICKSTUN (eds.)
49. Finite geometries and designs, P. CAMERON, J.W.P. HIRSCHFELD & D.R. HUGHES (eds.)
50. Commutator Calculus and groups of homotopy classes, H.J. BAUES
51. Synthetic differential geometry, A. KOCK
52. Combinatorics, H.N.V. TEMPERLEY (ed.)
53. Singularity theory, V.I. ARNOLD
54. Markov processes and related problems of analysis, E.B. DYNKIN
55. Ordered permutation groups, A.M.W. GLASS
56. Journees arithmetiques 1980, J.V. ARMITAGE (ed.)
57. Techniques of geometric topology, R.A. FENN
58. Singularities of differentiable functions, J. MARTINET
59. Applicable differential geometry, F.A.E. PIRANI and M. CRAMPIN
60. Integrable systems, S.P. NOVIKOV et al.

61. The core model, A. DODD
62. Economics for mathematicians, J.W.S. CASSELS
63. Continuous semigroups in Banach algebras, A.M. SINCLAIR
64. Basic concepts of enriched category theory, G.M. KELLY
65. Several complex variables and complex manifolds I, M.J. FIELD
66. Several complex variables and complex manifolds II, M.J. FIELD
67. Classification problems in ergodic theory, W. PARRY & S. TUNCEL

London Mathematical Society Lecture Note Series. 63

Continuous Semigroups in Banach Algebras

ALLAN M. SINCLAIR
Reader in Mathematics
University of Edinburgh

CAMBRIDGE UNIVERSITY PRESS

CAMBRIDGE

LONDON NEW YORK NEW ROCHELLE

MELBOURNE SYDNEY

Published by the Press Syndicate of the University of Cambridge
The Pitt Building, Trumpington Street, Cambridge CB2 1RP
32 East 57th Street, New York, NY 10022, USA
296 Beaconsfield Parade, Middle Park, Melbourne 3206, Australia

First published in 1982

Printed in Great Britain at the University Press, Cambridge

Library of Congress catalogue card number 81-21627

British Library cataloguing in publication data

Sinclair, Allan M.
Continuous semigroups in Banach algebras.
-(London Mathematical Society lecture note
series, ISSN 0076-0552; 63)
1. Banach algebras
I. Title II. Series
512'.55 QA326

ISBN 0 521 28598 4

CONTENTS

1 INTRODUCTION AND PRELIMINARIES

1.1 INTRODUCTION

The theory of analytic (one parameter) semigroups $t \mapsto a^t$ from the open right half plane H into a Banach algebra is the main topic discussed in these notes. Several concrete elementary classical examples of such semigroups are defined, a general method of constructing such semigroups in a Banach algebra with a bounded approximate identity is given, and then relationships between the semigroup and the algebra are investigated. These notes form small sections in the theory of (one parameter) continuous semigroups and in the general theory of Banach algebras. They emphasize an approach that is standard to neither of these subjects. A study of Hille and Phillips [1974] reveals that the theory of Banach algebras has been used as a tool in the study of certain problems in continuous semigroups, but that semigroup theory has until recently (1979) not impinged on the theory of Banach algebras. These lecture notes are about this recent progress.

Throughout these notes we use 'semigroup' for 'one parameter semigroup' when discussing a homomorphism from an additive subsemigroup of $\mathbb{C}$ into a Banach algebra, and we write our semigroups $t \mapsto a^t$ to emphasize the power law $a^{t+s} = a^t.a^s$ and function property of the semigroup. In the standard works on semigroups much attention is given to strongly continuous semigroups and their generators (see Hille and Phillips [1974], Dunford and Schwartz [1958], and Reed and Simon [1972]). In these works the generator itself is important, plays a fundamental role, and is often an object of considerable mathematical interest (for example, it may be the Laplacian). As the theory is developed here the generator is useful only in Chapter 6, and even there it is the resolvent $(1 - R)^{-1}$, not the generator R, that occurs in our Banach algebra results. It is possible in the Banach algebra situation to develop lemmas corresponding to the Hille-Yoshida Theorem totally avoiding unbounded closed operators and working with what is essentially the inverse of the generator. This seemed artificial and we do not do it here. In the standard works on semigroups (ibid.) most of the emphasis is on semigroups that are not quasinilpotent, and there is little or no space devoted to quasinilpotent semigroups (see Hille and Phillips [1974], p.481). However Chapters 5 and 6 of these notes concern radical Banach algebras, perhaps indirectly. In these radical algebras we are studying quasinilpotent semigroups.

The general theory of Banach algebras has mostly been developed for (Jacobson) semisimple algebras, and the most studied families of Banach algebras are semisimple: C*-algebras, group algebras, and uniform algebras. A brief glance through the standard references (Rickart [1960] and Bonsall and Duncan [1973]) illustrates this. Radical algebras and quasinilpotent elements play a very important role in Chapters 5 and 6 of these notes. However we do not attempt a study of radical Banach algebras or even discuss the role of non-continuous semigroups in the classification of radical Banach algebras. Various weaker assumptions on the domain of a semigroup $t \mapsto a^t$, for example, to the rational numbers, are related to the structure of certain radical Banach algebras (see Esterle [1980b]). Strongly continuous (one parameter) groups of automorphisms on a C*-algebra are fundamental in C*-algebra theory (see Pedersen [1979]). Except for this there had been few applications of semigroup theory to Banach algebras until 1979.

The standard references on the theory of semigroups (Hille and Phillips [1974], Dunford and Schwartz [1958], and Reed and Simon [1972]) contain much of Chapter 2 and the Hille-Yoshida Theorem of Chapter 6. The approach here is also basically different from that in Butzer and Berhens [1967], and Berge and Forst [1975]. The modification of the Cohen factorization theorem discussed in Chapters 3 and 4 is covered in considerable detail in Doran's and Wichman's lecture notes [1979] on bounded approximate identities and Cohen factorization. Even here our account differs from the original version, which is what they give.

These notes are elementary and the results are proved in detail. As background for the main results we assume standard elementary functional analysis, the complex analysis in <u>Real and Complex Analysis</u> by Rudin [1966], and the Banach algebra theory in <u>Complete Normed Algebras</u> by Bonsall and Duncan [1973]. We shall use the Titchmarsh convolution theorem (see Mikusiński [1959], Chapter 2) a couple of times. In a few corollaries and applications considerably more is assumed (for example, there are results applying to $L^1(G)$). Calculations are given in detail even when standard functions in $L^1(\mathbb{R})$ are being considered. The main tools in our proofs are techniques from Banach algebra theory and semigroup theory, the Bochner integral, and some classical results of complex analysis. Although the Hille-Yoshida and Ahlfors-Heins Theorems are standard results readily available in books, they are not in the assumed background and so they are proved in suitable forms in these notes (Theorem 6.7 and Appendix A1.1). In the introduction the Bochner integral is briefly discussed.

The notes are not polished. Each chapter beyond the first ends with notes and remarks where brief reference will be made to the literature, related results, and open problems. The bibliography is not comprehensive.

These notes are an expanded and revised version of lectures that I gave at the University of Edinburgh in January, February, and March 1980. The lectures and notes were both influenced by a course that J. Esterle gave in the University of California, Los Angeles, in April, May, and June 1979. Some parts of my lectures appear as they were given, others have been extensively revised, and occasionally a single verbal remark in a lecture has become a whole section here. The concrete semigroups in $L^1(\mathbb{R}^+)$ and $L^1(\mathbb{R}^n)$ were covered as here (Chapter 2) as was the Wiener Tauberian Theorem, (Theorem 5.6), Theorem 5.3, and the whole of Chapter 6. Chapters 3 and 4 were a single unproved result in lectures, but several of the audience had suffered talks from me on these subjects in a seminar.

I am grateful to many mathematicians for preprints and odd half forgotten conversations, which have influenced the development, and to the audience who survived my lectures. I am grateful to P.C. Curtis, Jr. and F.F. Bonsall for encouragement, to T.A. Gillespie for useful criticism of an early draft, to S. Grabiner for many discussions about Banach algebras, and to A.M. Davie for suggesting several improvements to results and proofs. H.G. Dales read the complete notes, and his detailed and careful criticism has enabled me to correct several errors and improve the notes. I am indebted to him for this and other suggestions. During 1978-9 J. Esterle and I had many discussions about radical Banach algebras and semigroups, and his U.C.L.A. lectures and seminars influenced my ideas. He has kindly given permission for me to include his results on nilpotent semigroups in Chapter 6 before he has published them. I am very grateful and deeply indebted to J. Esterle. Without his results in Chapters 5 and 6 these notes would not exist.

1.2 DEFINITIONS AND NOTATION

We shall now give some definitions, fix various notations, and prove a couple of useful little lemmas. Throughout these notes we shall consider complex Banach spaces and Banach algebras, and linear operators will be taken to be complex linear. The Banach algebras will not be assumed to have an identity, and these notes deal mainly with algebras without identity. If A is a Banach algebra, then $A \oplus \mathbb{C}1$ is the Banach algebra, obtained from A by formally adjoining an identity; note that the norm is $\|a + \lambda\| = \|a\| + |\lambda|$ for all $a \in A$ and $\lambda \in \mathbb{C}$. If A is a Banach algebra with identity $A^{\#} = A$, and if A is an algebra without identity $A^{\#} = A \oplus \mathbb{C}1$. The algebra $A^{\#}$ is the algebra in which the spectra of elements of A are calculated. The spectrum of $x \in A$ is denoted by $\sigma(x)$ and the spectral radius by $\nu(x)$.

If f is a function from a set X into a set Y, we shall often write $x \mapsto f(x) : X \to Y$. If X is a Banach space, $BL(X)$ denotes the Banach algebra of bounded linear operators on X. For a commutative Banach algebra A the <u>multiplier algebra</u> $Mul(A)$ is defined to be the set of $T \in BL(A)$ such that $T(ax) = a\,T(x)$ for all $x, a \in A$. Clearly $Mul(A)$ is a unital Banach algebra, and there is a natural norm reducing homomorphism $a \mapsto L_a : A \to Mul(A)$, where $L_a x = ax$ for all $x \in A$. Let Ω be a locally compact Hausdorff space and let $C_o(\Omega)$ be the Banach algebra of continuous complex valued functions on Ω vanishing at infinity. Then $C_o(\Omega)^{\#}$ is isomorphic to $C(\Omega \cup \{\infty\})$, where $\Omega \cup \{\infty\}$ is the one point compactification of Ω, and $Mul(C_o(\Omega))$ is isomorphic to $C(\beta\Omega)$, where $\beta\Omega$ is the Stone-Čech compactification of Ω.

Most of the Banach algebras we study have bounded approximate identities. A Banach algebra A has a <u>bounded approximate identity</u> Λ bounded by d if $\|f\| \le d$ for all $f \in \Lambda$, and if, for each finite subset $F \subseteq A$ and each $\varepsilon > 0$, there is an $e \in \Lambda$ such that $\|ea - a\| + \|ae - a\| < \varepsilon$ for all $a \in F$. If the set Λ can be chosen to be countable (commutative,...), we say that A has a countable (commutative,...) bounded approximate identity. If $\Lambda = \{a \in A : \|a\| \le d\}$ we shall suppress Λ. In Chapter 3 the countability of the bounded approximate identity is important. Here we note a couple of folklore facts which indicate that this hypothesis is not too restrictive for our purposes.

If a Banach algebra A is separable and has a bounded approximate identity, then A has a countable bounded approximate identity. This can be seen by choosing a countable dense subset $\{y_n\}$ of A, and then choosing a sequence (e_n) from the bounded approximate identity such that $\| e_n y_j - y_j \| + \| y_j e_n - y_j \| < n^{-1}$ for $1 \le j \le n$ and all n. The set $\{e_n : n \in \mathbb{N}\}$ is a countable bounded approximate identity in A.

If A is a Banach algebra with a bounded approximate identity and if Y is a separable subspace of A, then there is a separable Banach subalgebra B of A that contains Y and has a bounded approximate identity. Let $\{y_n\}$ be a countable dense subset of Y and choose a sequence (e_n) from the bounded approximate identity of A such that
$\| e_n y_j - y_j \| + \| y_j e_n - y_j \| < n^{-1}$ and $\| e_n e_j - e_j \| + \| e_j e_n - e_j \| < n^{-1}$
for $1 \le j \le n - 1$ and all n. The Banach subalgebra B of A generated by $Y \cup \{e_n : n \in \mathbb{N}\}$ has the required properties.

If a commutative Banach algebra A has a bounded approximate identity bounded by 1, then the natural homomorphism $a \mapsto L_a : A \to \mathrm{Mul}(A)$ from A into the multiplier algebra is an isometric embedding from A onto a closed ideal in $\mathrm{Mul}(A)$.

The complex numbers, real numbers, integers, and positive integers are denoted by $\mathbb{C}$, $\mathbb{R}$, $\mathbb{Z}$, and $\mathbb{N}$, respectively. The open right half plane $\{z \in \mathbb{C} : \mathrm{Re}\, z > o\}$ is denoted by H, and the closed right half plane by H^-. The reader will be reminded of this notation periodically.

A function f from an open subset U of $\mathbb{C}$ into a Banach space X is said to be <u>analytic</u> if for each $z \in U$ the limit $\lim_{h \to o} h^{-1}(f(z+h) - f(z))$ exists in X. This limit is denoted by $(Df)(z)$.
The Hahn-Banach Theorem may be combined with results of complex analysis to yield results about analytic functions into a Banach space. The following result illustrates this technique and shows why it is not necessary to consider separately semigroups which are weakly or strongly analytic.

1.3 <u>LEMMA</u>

Let f be a function from an open subset U of the complex plane into a Banach space X. Then conditions (a), (b), and (c) on f are equivalent.

(a) f is analytic.

(b) Ff is analytic for all $F \in X^*$

(c) f is continuous, and

$$f(z) = \frac{1}{2\pi i} \int_\gamma f(\xi)\ (\xi - z)^{-1}\, d\,\xi,$$

for each $z \in U$ and each closed path γ in U such that the winding number of z with respect to γ is 1 and of each point in $\mathbb{C} \setminus U$ is 0.

(d) If $X = BL(Y)$ for a Banach space Y, then the above conditions are equivalent to $z \mapsto F(f(z)y) : U \to \mathbb{C}$ being analytic for all $y \in Y$ and $F \in Y^*$.

Proof Clearly (a) implies (b) and (d).

If (b) holds and if $z, z_n \in U$ with $z_n \to z$, then $F((z_n - z)^{-1}(f(z_n) - f(z)))$ converges in $\mathbb{C}$ for each $F \in X^*$. The uniform boundedness theorem implies that the sequence $(\|(z_n - z)^{-1}(f(z_n) - f(z))\|)$ is bounded. Thus f is continuous on U. The integral in (c) now converges in X (see 1.6), and using the classical Cauchy integral formula for a complex valued analytic function we obtain

$$F(f(z)) = \frac{1}{2\pi i} \int_\gamma (Ff)(\xi)(\xi - z)^{-1}\, d\xi$$

$$= F\left(\frac{1}{2\pi i} \int_\gamma f(\xi)(\xi - z)^{-1}\, d\xi\right)$$

for all $F \in X^*$. An application of the Hahn-Banach Theorem gives the equality of (c).

Now suppose that (c) holds. Let $z \in U$ and let γ be a small circle with centre z and radius r so that γ and its interior are contained in U. Let $M = \sup\{\|f(\xi)\| : \xi \in \gamma\}$. If $h \in \mathbb{C}$ with $|h| < r/2$, then

$$f(z + h) - f(z) - \frac{h}{2\pi i} \int_\gamma f(\xi)(\xi - z)^{-2}\, d\xi$$

$$= \frac{h^2}{2\pi i} \int_\gamma f(\xi)\ (\xi - z)^{-2}\ (\xi - z - h)^{-1}\, d\xi$$

so that

$$\left\| f(z+h) - f(z) - \frac{h}{2\pi i} \int_\gamma f(\xi)(\xi - z)^{-2}\, d\xi \right\|$$

$$\leq \frac{|h|^2}{2\pi} \cdot 2\pi r \cdot \frac{1}{r^2} \cdot \frac{1}{r/2} \cdot M.$$

This shows that f has derivative $\frac{1}{2\pi i} \int_\gamma f(\xi)\ (\xi - z)^{-2}\, d\xi$ at z, as we would expect from classical complex analysis. Hence (a) holds.

Suppose that (d) holds. Using the equivalence of (a) and (b) we find that $z \mapsto f(z)y : U \to Y$ is analytic for each $y \in Y$. An application of the uniform boundedness theorem as above shows that f is continuous on U. The integral in (c) now exists, and the equality in (c) is obtained by an application of the Hahn-Banach Theorem. The proof is complete.

A (one-parameter) <u>semigroup</u> in a Banach algebra A is a function $t \mapsto a^t$ from an additive subsemigroup of $\mathbb{C}$ containing the open right half line $(0,\infty)$ into A satisfying $a^{t+r} = a^t . a^r$ for all t,r in the domain of definition. We shall be concerned mainly with semigroups defined on the open right half plane H and on the open right half line $(0,\infty)$. The semigroup is said to be analytic or continuous (in some topology on A) if the function is analytic or continuous. Lemma 1.3 shows why we restrict attention to the norm topology when considering analytic semigroups defined on H. A semigroup $t \mapsto a^t$ is said to be a <u>contraction</u> semigroup if $\|a^t\| \leq 1$ for all t.

1.4 <u>LEMMA</u>

Let $t \mapsto a^t : H \to A$ be an analytic semigroup from the open right half plane into a Banach algebra A. Then $(a^t A)^- = (a^1 A)^-$ and $(A\, a^t)^- = (A\, a^1)^-$ for all $t \in H$.

<u>Proof</u>. Let $F \in A^*$ with $F(a^1 A) = \{0\}$. Then $t \mapsto F(a^t x) : H \to \mathbb{C}$ is an analytic function for each $x \in A$. This function is zero for all $t \in \mathbb{C}$ with $\operatorname{Re} t > 1$ because $F(a^t x) = F(a^1 a^{t-1} x) = 0$. Hence $F(a^t x) = 0$ for all $t \in H$ and $x \in A$. By the Hahn-Banach Theorem it follows that $(a^t A)^- \subseteq (a^1 A)^-$, and a similar argument yields the reverse inclusion.

Semigroups $t \mapsto a^t : (0,\infty) \to A$ that are not analytic may have

$(a^t A)^-$ strictly decreasing, and Chapter 6 is devoted to the study of such semigroups. However even there our aim is to study continuous rather than strongly continuous semigroups.

1.5 AN OUTLINE OF THE NOTES

This section is a synopsis of the notes. In Chapter 2 examples of analytic semigroups from the open right half plane into several concrete Banach algebras are discussed. These semigroups will suggest abstract properties to be investigated in Chapter 3 and will provide examples for the results and proofs of Chapter 5. The structure of C^*-algebras is very rich, and this makes it easy to construct analytic semigroups in C^*-algebras by using the commutative Gelfand-Naimark Theorem to define a^t for a positive element in the algebra and t in H. However these semigroups in C^*-algebras are not useful tools to study the algebra. Semigroups in various convolution algebras are more interesting than in C^*-algebras and throw more light on the structure of the algebra. The fractional integral semigroup $t \mapsto I^t$, where $I^t(w) = \Gamma(t)^{-1} w^{t-1} e^{-w}$, and the backward heat semigroup $t \mapsto C^t$, where $C^t(w) = t(2\pi^{\frac{1}{2}})^{-1} w^{-3/2} \exp(-t^2/4w)$, are given in detail as examples of semigroups into $L^1(\mathbb{R}^+)$ (see Theorem 2.6, and Lemma 2.9).

The Banach algebra $L^1(\mathbb{R}^+)$ is very important in the study of semigroups in Banach algebras for the following reason. If $t \mapsto a^t : (0,\infty) \to A$ is a continuous contraction semigroup into a Banach algebra A, then there is a natural norm reducing homomorphism Θ from $L^1(\mathbb{R}^+)$ into A defined by $\Theta(f) = \int_0^\infty f(t)a^t\, dt$. The integral is a Bochner integral (1.6), the norm reducing property $(\|\Theta\| \le 1)$ follows from $\|a^t\| \le 1$ for all $t > 0$, and the homomorphism property follows by a change in the order of integration. The homomorphism Θ may be used to map analytic semigroups in $L^1(\mathbb{R}^+)$ into analytic semigroups in A. The image semigroups are called <u>subordinate</u> to the semigroup $t \mapsto a^t$. The <u>Gaussian</u> semigroup

$$G^t(w) = (4\pi t)^{-n/2} \exp(-|w|^2/4^t)$$

and <u>Poisson</u> semigroup

$$P^t(w) = \frac{\Gamma((n+1)/2)}{\pi^{(n+1)/2}} \cdot \frac{t}{(t^2 + |w|^2)^{(n+1)/2}}$$

in $L^1(\mathbb{R}^n)$ are discussed, and the Poisson semigroup is studied via its subordination to the Gaussian semigroup. Certain growth properties of $\|G^t\|_1$ are substantially better than those of $\|I^t\|_1$. These strong growth properties of $\|G^t\|_1$ are crucial in the proof of the Wiener Tauberian Theorem (Theorem 5.6).

Chapter 3 contains a theorem that gives the existence of analytic semigroups in a Banach algebra with a countable bounded approximate identity. The result is proved by modifying the proof of Cohen's factorization theorem so that the proof resembles the way in which a strongly continuous semigroup is generated in the Hille-Yoshida Theorem (see 6.7). The semigroup constructed in Chapter 3 has growth and structure more like the fractional integral semigroup than the Gaussian semigroup. The general semigroup result is applied to the group algebra $L^1(G)$ of a metrizable locally compact group, and to obtain commutative bounded approximate identities in Banach algebras with countable bounded approximate identities.

The proof of the main result in Chapter 3 and the lemmas required in the proof fill Chapter 4. Two of the properties (6 and 15) of Theorem 3.1 I have not been able to prove by the exponential methods of Chapter 4. These two properties require the factorization results developed in Sinclair [1979a], although I believe they may be obtained from exponential calculations. I have attempted to prove the most general factorization result for semigroups that I know. In other chapters generality has often been sacrificed to obtain an elementary account.

The properties of the semigroups near the boundaries of their domains of definition are interesting, and are closely related to the fine structure of the semigroup and the algebra. In Chapter 5 we investigate the restrictions imposed on a commutative Banach algebra A by the assumption that it contains an analytic semigroup $t \mapsto a^t : H \to A$ with $(a^1 A)^- = A$ such that the growth of $\|a^t\|$ is suitably restricted. The restrictions we consider are:

(i) to the growth of $\|a^t\|$ along rays in H emanating from 0 (5.2);

(ii) to the growth of $\|a^t\|$ along a vertical line (5.5);

(iii) the boundedness in the semidisc $\{z \in H : |z| \le 1\}$ (5.12).

In each of these cases we prove a result due to Esterle : these are, respectively, a result about radical Banach algebras, a Tauberian theorem, and a result on the non-separability of the multiplier algebra. The philosophy underlying the proofs in the first two cases is to define a suitable classical analytic function F (by using the analyticity of the semigroup and a continuous linear functional on the algebra), and to apply the Ahlfors-Heins Theorem to F.

Chapter 6 begins with a standard account of the Hille-Yoshida Theorem relating a strongly continuous contraction semigroup a^t on a Banach space with the closed operator R that is its infinitesimal generator. The relationships between the nilpotency of the semigroup and the growth of $\|(1 - R)^{-n}\|$ as n tends to infinity is investigated. This result is applied to a hyperinvariant subspace theorem for suitable operators on a Banach space, and to prove the existence of a proper closed ideal in a commutative radical Banach algebra containing a non-zero element u such that $\|u(\lambda - u)^{-1}\| \leq 1$ for all $\lambda > 0$ and $\{n\|u^n\|^{1/n} : n \in \mathbb{N}\}$ is bounded. This process is seen to give abstractly the obvious ideals in the Volterra convolution algebra $L^1_\star[0,1]$.

In the appendix we give a proof of a special case of the Ahlfors-Heins Theorem, and prove a theorem of G.R. Allan [1979] on certain closed ideals of $L^1(\mathbb{R}^+,\omega)$ for ω a radical weight. It is also shown that an Arens regular Banach algebra with a bounded approximate identity has a bounded approximate identity that is well behaved with respect to derivations (and is quasicentral).

1.6 INTEGRALS

We shall frequently integrate continuous functions on $(0,\infty)$ with values in a Banach space, and in this section we briefly define the elementary integrals used. There are extensive discussions of the integration of Banach space valued functions in Dunford and Schwartz [1958] and in Hille and Phillips [1974].

Let f be a continuous function from $(0,\infty)$ into a Banach space X with $\int_0^\infty \|f(w)\|\,dw$ finite. For $0 < \alpha < \beta < \infty$ we define $\int_\alpha^\beta f(w)\,dw$ as the limit of Riemann sums using partitions $\alpha = \alpha_0 < \alpha_1 < \ldots < \alpha_n = \beta$ with $\max\{\alpha_j - \alpha_{j-1} : 1 \leq j \leq n\}$ tending to zero. The limit may be shown to exist in X in the same way that the classical Riemann integral is shown to exist by using the uniform continuity of the integrand. Further the integral $\int_\alpha^\beta f(w)\,dw$ is seen to be linear in f, and to satisfy $\left\|\int_\alpha^\beta f(w)\,dw\right\| \leq \int_\alpha^\beta \|f(w)\|\,dw$ and $F\left(\int_\beta^\alpha f(w)\,dw\right) = \int_\alpha^\beta (Ff)(w)\,dw$ for all $F \in X^*$. Using the observation that $\int_0^\alpha + \int_\beta^\infty \|f(w)\|\,dw$ tends to zero as α tends to zero and β tends to infinity, we define $\int_0^\infty f(w)dw$ to be the limit of $\int_\alpha^\beta f(w)dw$. The integral $\int_0^\infty$ satisfies similar properties to $\int_\alpha^\beta$. Let $t \mapsto b(t) : (0,\infty) \to X$ be a continuous function into X with $\|b(t)\| \le M$ for all $t > 0$. The linear operator $g \mapsto \int_0^\infty g(t)\, b(t)dt$ from the space $(C_0(0,\infty), \|.\|_1)$ of continuous complex valued functions on $(0,\infty)$ vanishing at 0 and ∞ with the $\|.\|_1$-norm into X is continuous. Using the density of $C_0(0,\infty)$ in $L^1(\mathbb{R}^+)$ the integral $\int_0^\infty g(t)\, b(t)\, dt$ may be defined for all $g \in L^1(\mathbb{R}^+)$ by approximating g by continuous functions. Further $\|\int_0^\infty g(t)\, b(t)\, dt\| \le M\, \|g\|_1$ and $F(\int_0^\infty g(t)\, b(t)\, dt) = \int_0^\infty g(t)\, (Fb)\, (t)\, dt$ for all $g \in L^1(\mathbb{R}^+)$ and all $F \in X^*$, where the last integral is a Lebesgue integral. If $t \mapsto T(t) : (0,\infty) \to BL(X)$ is a strongly continuous function from $(0,\infty)$ into the Banach space of bounded linear operators on X with $\int_0^\infty \|T(t)\|\, dt$ finite, then we define $\int_0^\infty T(t)dt$ as an operator in $BL(X)$ by $(\int_0^\infty T(t)dt)\, x = \int_0^\infty T(t)\, x\, dt$ for all $x \in X$. Note that

$$\|\int_0^\infty T(t)\, dt\| \le \int_0^\infty \|T(t)\|\, dt.$$

2 ANALYTIC SEMIGROUPS IN PARTICULAR BANACH ALGEBRAS

2.1 INTRODUCTION

In this chapter we introduce various well known semigroups from the open right half plane H into particular Banach algebras. We discuss the power semigroups in a separable C^*-algebra, the fractional integral and backwards heat semigroups in $L^1(\mathbb{R}^+)$, and the Gaussian and Poisson semigroups in $L^1(\mathbb{R}^n)$. While doing this we shall develop notation that is used in subsequent chapters. The discussion is very detailed throughout the chapter, and is designed to introduce and motivate following chapters dealing with more abstract results for analytic semigroups. For example we are concerned with the asymptotic behaviour of $\|a^{1+iy}\|$ as $|y|$ tends to infinity, but not with the infinitesimal generators of our semigroups even though they are important. We shall discuss generators in a different context in Chapter 6.

2.2 C^*-ALGEBRAS

The functional calculus for a positive hermitian element in a C^*-algebra that is derived from the commutative Gelfand-Naimark Theorem enables us to construct very well behaved semigroups in C^*-algebras. We shall briefly discuss the case of a commutative C^*-algebra before we state and prove our main result on semigroups in a C^*-algebra. The commutative Gelfand-Naimark Theorem (see, for example, Bonsall and Duncan [1973]) enables us to identify the commutative C^*-algebra with $C_o(\Omega)$, which is the C^*-algebra of continuous complex valued functions vanishing at infinity on the locally compact Hausdorff space Ω. It is easy to check that $C_o(\Omega)$ has a countable bounded approximate identity if and only if Ω is σ-compact (that is, Ω is a countable union of compact subsets of itself). By using a countable bounded approximate identity in $C_o(\Omega)$, or by using the σ-compactness of Ω, an $f \in C(\Omega)$ may be constructed so that $1 \geq f(\phi) > 0$ for all $\phi \in \Omega$. The analytic semigroup $t \mapsto f^t : H \to C_o(\Omega)$

is given by defining $f^t(\phi) = f(\phi)^t$ for all $\phi \in \Omega$ and $t \in H$.

2.3 THEOREM

A C^*-algebra A has a countable bounded approximate identity if and only if there is an analytic semigroup $t \mapsto a^t : H \to A$ such that $(a^tA)^- = A = (Aa^t)^-$ and $\|a^t\| \le 1$ for all $t \in H$, $a^t \ge 0$ for all $t > 0$, and $\|a^t x - x\| + \|xa^t - x\| \to 0$ as $t \to 0$ in H for all $x \in A$.

Proof. If A contains a semigroup with the required properties, then $\{a^{1/n} : n \in \mathbb{N}\}$ is a countable bounded approximate identity in A. Conversely suppose that $\{g_n : n \in \mathbb{N}\}$ is a countable bounded approximate identity in A. For each n let $e_n = g_n g_n^*$. To show that $\{e_n : n \in \mathbb{N}\}$ is a bounded approximate identity in A, it is sufficient to show that $\|x(e_n - 1)\| \to 0$ as $n \to \infty$ for all $x \in A$ because $\|(e_n - 1)x\| = \|x^*(e_n - 1)\|$.
Now

$$\|x(e_n - 1)\|$$

$$\le \|x(g_n - 1)\| + \|xg_n(g_n^* - 1)\|$$

$$\le \|x(g_n - 1)\| + \|x(g_n - 1)\| (\|g_n^*\| + 1) + \|x(g_n^* - 1)\|,$$

and $\|x(g_n^* - 1)\| = \|(g_n - 1)x^*\|$ for all $n \in \mathbb{N}$ and $x \in A$. Hence $\|x(e_n - 1)\|$ tends to 0 as N tends to infinity.

Let $a = \sum_{j=1}^{\infty} e_j^2 2^{-j} \|e_j\|^{-2}$. Then $0 \le a$ and $\|a\| \le 1$. We apply the Gelfand-Naimark Theorem for a commutative C^*-algebra to the C^*-algebra generated by a. This gives a norm reducing $*$-homomorphism θ from the C^*-algebra $\{f \in C[0,1] : f(0) = 0\}$ onto the commutative C^*-algebra generated by a with $\theta(z) = a$. We take $a^t = \theta(z^t)$, where $z^t(w) = w^t$ for all $w \in [0,1]$ and all $t \in H$. From the definition of a^t and properties of semigroups in $\mathbb{C}$, we observe that $t \mapsto a^t : H \to A$ is an analytic semigroup such that $\|a^t\| \le 1$ for all $t \in H$, $a^t \ge 0$ for all $t > 0$, and $\|a^t a - a\| \to 0$ as $t \to 0$, $t \in H$. If we show that $\|e_n a^t - e_n\|$ tends to zero as t tends to zero, $t \in H$, then we shall have completed the proof for the following reason. If $t \in H$, then $\bar{t} \in H$ where $\bar{t}$ is the complex conjugate of t and $\|a^t e_n - e_n\| = \|e_n (a^t)^* - e_n\| = \|e_n a^{\bar{t}} - e_n\|$. Also $\{e_n : n \in \mathbb{N}\}$ is a bounded approximate identity for A. Thus the closures of $\{e_n A : n \in \mathbb{N}\}$ and $\{Ae_n : n \in \mathbb{N}\}$ are both equal to A. To prove that

$\|e_n a^t - e_n\|$ tends to zero as t tends to zero, $t \in H$, we require the following standard little lemma on C^*-algebras.

2.4 LEMMA

Let x, y, b be in a C^*-algebra A with $0 \leq x$ and $0 \leq y$. If $0 \leq x^2 \leq y^2$, then $\|xb\| \leq \|yb\|$.

Proof. If f is a positive linear functional on A, then $z \mapsto f(b^* zb) : A \to \mathbb{C}$ is a positive linear functional on A since $f(b^* z^* zb) = f((zb)^* zb) \geq 0$. Hence $f(b^* y^2 b) \geq f(b^* x^2 b)$ for all positive linear functionals f on A. Since the norm of a positive element c in a C^*-algebra A is $\sup\{f(c) : f \geq 0, \|f\| \leq 1\}$, we have $\|b^* y^2 b\| \geq \|b^* x^2 b\|$ so $\|yb\|^2 = \|(yb)^* yb\| \geq \|(xb)^* xb\| = \|xb\|^2$. This proves the lemma.

We apply the lemma in the C^*-algebra obtained by adjoining an identity to A. Let $b = 1 - a^t$, $x = e_n . 2^{-n/2} \|e_n\|^{-1}$, and $y = a^{1/2}$. Then $\|e_n(1 - a^t)\| \leq 2^{n/2} \|e_n\| . \|a(1 - a^t)\|$ for all $t \in H$, and the proof is complete.

The proof of Theorem 2.3 shows that a C^*-algebra with a countable bounded approximate identity has a commutative bounded approximate identity.

2.5 THE CONVOLUTION ALGEBRA $L^1(\mathbb{R}^+)$

In Corollary 3.5 we shall see that there is a norm reducing homomorphism θ from $L^1(\mathbb{R}^+)$ into a Banach algebra A with a countable bounded approximate identity bounded by 1. Under this homomorphism (analytic) semigroups in $L^1(\mathbb{R}^+)$ are mapped into (analytic) semigroups in the algebra A. This is an old idea first systematically exploited by Bochner [1955] and called the subordination of one semigroup to another. We shall subsequently use the homomorphism θ from $L^1(\mathbb{R}^+)$ into A several times. The growth estimates on semigroups in $L^1(\mathbb{R}^+)$ will hold for some semigroups in A.

Recall that $L^1(\mathbb{R}^+)$ is the Banach space of (equivalence classes of) integrable functions f on $\mathbb{R}^+ = [0,\infty)$ with norm $\|f\|_1 = \int_{\mathbb{R}^+} |f(w)| dw$. With the convolution product

$$(f*g)(t) = \int_0^t f(t-w)\, g(w)\, dw$$

defined almost everywhere, $L^1(\mathbb{R}^+)$ is a commutative Banach algebra without identity. The Banach algebra $L^1(\mathbb{R}^+)$ is semisimple and its carrier space Φ may be identified with the closed right half plane H^- by the mapping $\lambda \mapsto \phi_\lambda : H^- \to \Phi$ where ϕ_λ is defined by

$$\phi_\lambda(f) = \int_{\mathbb{R}^+} f(w)\, e^{-\lambda w}\, dw$$

for all $f \in L^1(\mathbb{R}^+)$. The Gelfand map $f \mapsto \hat{f} : L^1(\mathbb{R}^+) \to C_o(\Phi)$ is just the Laplace transform L, where $(Lf)(\lambda) = \int_{\mathbb{R}^+} f(w) e^{-\lambda w}\, dw$, because $\hat{f}(\phi_\lambda) = \phi_\lambda(f) = (Lf)(\lambda)$. The Laplace transform Lf is continuous on H^- and analytic in H. The algebra $L^1(\mathbb{R}^+)$ has a countable bounded approximate identity bounded by 1 : for example, $\{e_n : n \in \mathbb{N}\}$, where

$$e_n(w) = \begin{cases} n & \text{for } 0 \le w \le 1/n \\ 0 & \text{for } 1/n < w \end{cases}$$

The Titchmarsh convolution theorem (or properties of analytic functions via the Laplace transform) implies that $L^1(\mathbb{R}^+)$ is an integral domain; that is, if $f, g \in L^1(\mathbb{R}^+)$ with $f*g = 0$ then $f = 0$ or $g = 0$. The multiplier algebra $Mul(L^1(\mathbb{R}^+))$ of $L^1(\mathbb{R}^+)$ is naturally isometrically isomorphic with the convolution measure algebra $M(\mathbb{R}^+)$ on $\mathbb{R}^+$. Here $M(\mathbb{R}^+)$ is the Banach space of bounded regular Bonel measures μ on $\mathbb{R}^+$ with norm $\|\mu\| = |\mu|(\mathbb{R}^+)$ and product $\mu * \nu$ defined by $(\mu*\nu)(E) = \iint \chi_E(w+u)\, d\mu(w)\, d\nu(u)$, where χ_E is the characteristic function of the Borel set E. The isometric isomorphism from $M(\mathbb{R}^+)$ onto $Mul(L^1(\mathbb{R}^+))$ is defined by $\mu \mapsto T_\mu : M(\mathbb{R}^+) \to Mul(L^1(\mathbb{R}^+))$, where $T_\mu f = \mu*f$ (see Johnson [1964]).

Recall that the Gamma function Γ is defined by

$$\Gamma(t) = \int_0^\infty w^{t-1}\, e^{-w}\, dw$$

for all $t \in H$, and that asymptotically $\Gamma(t) = t^t(2\pi/t)^{1/2} \exp(-t + O(|t|^{-1}))$ as $|t| \to \infty$ for all $t \in H$ (see Erdelyi [1953, p.21] or Olver [1974, p.294]). Further Γ is analytic in H, $\Gamma(t+1) = t\Gamma(t)$ for all $t \in H$, $\Gamma(n+1) = n!$ for all $n \in \mathbb{N}$, and

$\Gamma(1/2) = \pi^{1/2}$

2.6 THEOREM

If $I^t \in L^1(\mathbb{R}^+)$ is defined by $I^t(w) = w^{t-1} e^{-w} \Gamma(t)^{-1}$ for all $w \in (0,\infty)$ and all $t \in H$, then $t \mapsto I^t : H \to L^1(\mathbb{R}^+)$ is an analytic semigroup, called the fractional integral semigroup, with the following properties.

(i) $(I^t * L^1(\mathbb{R}^+))^- = L^1(\mathbb{R}^+)$ for all $t \in H$.

(ii) $\|I^{x+iy}\|_1 = \dfrac{\Gamma(x)}{|\Gamma(x+iy)|}$ for all $x > 0$ and all $y \in \mathbb{R}$.

(iii) $(\mathcal{L}I^t)(z) = (z+1)^{-t}$ for all $z \in H^-$ and all $t \in H$, and $\sigma(I^t) = \{0\} \cup \{(z+1)^{-t} : z \in H^-\}$ for all $t \in H$.

(iv) If $x+iy \in H$, then

$$\|I^{x+iy}\|_1 = K(x)(1 + y^2x^{-2})^{-\frac{x}{2}+\frac{1}{4}} . \exp\left(\frac{\pi|y|}{2} + O(|y|^{-1}) \right)$$

where $K(x)$ is a constant depending on x but not on y, and O is independent of x.

Before proving Theorem 2.6 we prove two little lemmas, which will be useful in proving the analyticity of the semigroup and in checking condition (i). We shall use these lemmas several times in this chapter.

2.7 LEMMA

Let (W, Σ, μ) be a measure space with μ a positive measure, let $1 \le p \le \infty$, and let $(t,w) \mapsto F(t,w) : H \times W \to \mathbb{C}$ be a measurable function such that $w \mapsto F(t,w)$ is in $L^p(W)$ for each $t \in H$ and $t \mapsto \|F(t,\cdot)\|_p : H \to \mathbb{R}$ is continuous. If $t \mapsto F(t,w) : H \to \mathbb{C}$ is analytic for each $w \in W$, then $t \mapsto F(t,\cdot) : H \to L^p(W)$ is analytic.

Proof of Lemma. Let $t = x+iy \in H$, let $0 < r < x/2$, and let C be the circle with centre 0 and radius r. By Cauchy's integral formula,

$$F(t+h, w) - F(t,w) - h\frac{\partial F(t,w)}{\partial t} = \frac{h^2}{2\pi i}\int_C \frac{F(t+z, w)}{z^2(z-h)}\,dz$$

so that

$$|F(t+h, w) - F(t,w) - h\frac{\partial F(t,w)}{\partial t}| \le \frac{|h|^2}{\pi r^2}\int_0^{2\pi} |F(t+re^{i\theta}, w)|\,d\theta$$

for all $w \in W$ and $h \in \mathbb{C}$ with $|h| < r/2$. Since $w \mapsto \frac{\partial F}{\partial t}(t,w) : W \to \mathbb{C}$ is a pointwise limit of a sequence of measurable functions, it is measurable. Raising the last inequality to the power p and integrating over W we obtain

$$\|F(t+h,\cdot) - F(t,\cdot) - h\frac{\partial F}{\partial t}(t,\cdot)\|_p$$
$$\le \frac{|h|^2}{\pi r^2}\left\{\int\left(\int_0^{2\pi}|F(t+re^{i\theta},w)|\,d\theta\right)^p d\mu(w)\right\}^{1/p}$$

and using Hölder's Inequality on the $d\theta$ integral

$$\le \frac{|h|^2}{\pi r^2}\left\{\int\int_0^{2\pi}|F(t+re^{i\theta},w)|^p\,d\theta.(2\pi)^{(1-\frac{1}{p})p}\,d\mu(w)\right\}^{1/p}$$
$$\le \frac{|h|^2(2\pi)^{1-1/p}}{\pi r^2}\left\{\int_0^{2\pi}\|F(t+re^{i\theta},\cdot)\|_p^p\,d\theta\right\}^{1/p}$$

by Fubini's Theorem. We have proved the case $1 < p < \infty$ and the cases $p = 1$ and $p = \infty$ are similar but do not require Hölder's inequality. Hence $\frac{\partial F}{\partial t}(t,\cdot) \in L^p(W)$ and $t \mapsto F(t,\cdot) : H \to L^p(W)$ is analytic. This completes the proof.

A standard argument involving bounded approximate identities in convolution algebras is used in the proof of Lemma 2.8.

2.8 <u>LEMMA</u>

Let A be one of the convolution algebras $L^1(\mathbb{R}^+)$ or $L^1(\mathbb{R}^n)$, and let $t \mapsto a^t : H \to A$ be an analytic semigroup in A with $a^t \ge 0$ almost everywhere and $\|a^t\|_1 = 1$ for all $t > 0$. Then $(a^t * A)^- = A$ for all $t \in H$ if and only if $\int_{|w|\ge\delta} a^t(w)\,dw \to 0$ as $t \to 0$, $t > 0$, for all $\delta > 0$.

<u>Proof</u>. We shall only use and prove the if implication of this lemma. By Lemma 1.4 to prove that $(a^t*A)^- = A$ for all $t \in H$, it is sufficient to show that $\|a^t*g - g\|_1 \to 0$ as $t \to 0$, $t > 0$, for each $g \in A$. Since $\|a^t\|_1 = 1$ for all $t > 0$ and since the set of continuous functions with compact supports is dense in A, we may assume that g is continuous with compact support. The function g is then uniformly continuous so for $\varepsilon > 0$

there is a $1 > \delta > 0$ such that $|g(u) - g(w)| < \varepsilon$ for all u,w with $|u - w| < \delta$. Let C be the sum of the support of g and the closed unit sphere in $\mathbb{R}^n$ or $\mathbb{R}^+$ (in the latter case the sphere is $[0,1]$). If $t > 0$, then

$$\begin{aligned}
&\|a^t * g - g\|_1 \\
&= \int \left| \int (g(w - u) - g(w))\, a^t(u)\, du \right| dw \\
&\le \int \int_{|u|<\delta} |g(w-u) - g(w)|\, a^t(u)\, du\, dw \\
&+ \int \int_{|u|\ge\delta} |g(w-u) - g(w)|\, a^t(u)\, du\, dw \\
&\le \int_C \int_{|u|<\delta} \varepsilon\, a^t(u)\, du\, dw + \int_{|u|\ge\delta} 2\,\|g\|_1\, a^t(u)\, du \\
&\le \varepsilon \int_C dw + 2\|g\|_1 \int_{|u|\ge\delta} a^t(u)\, du.
\end{aligned}$$

Choosing ε small enough and then letting $t \to 0$, $t > 0$, completes the proof.

Proof of Theorem 2.6. If $t = x + iy \in H$ with x and y real, then I^t is a continuous function on $(0,\infty)$, and

$$\|I^t\|_1 = \int_0^\infty \frac{w^{x-1}\, e^{-w}}{|\Gamma(x+iy)|}\, dw = \frac{\Gamma(x)}{|\Gamma(x+iy)|}.$$

Thus $I^t \in L^1(\mathbb{R}^+)$ for all $t \in H$, and $t \mapsto \|I^t\|_1 : H \to \mathbb{R}$ is continuous. Further $t \mapsto I^t(w) : H \to \mathbb{C}$ is analytic and

$$\frac{\partial I^t}{\partial t}(w) = \frac{w^{t-1}\, e^{-w}}{\Gamma(t)} \left\{ \log w - \frac{\Gamma'(t)}{\Gamma(t)} \right\}$$

for each $w > 0$. By Lemma 2.7, $t \mapsto I^t : H \to L^1(\mathbb{R}^+)$ is an analytic function. If $s,t,w > 0$, then

$$\begin{aligned}
&I^t * I^s(w) \\
&= \int_0^w \frac{(w-u)^{t-1}\, e^{-(w-u)}\, u^{s-1}\, e^{-u}}{\Gamma(t).\Gamma(s)}\, du \\
&= \frac{w^{t+s-1}\, e^{-w}}{\Gamma(t)\ \Gamma(s)} \int_0^1 (1 - v)^{t-1}\, v^{s-1}\, dv,
\end{aligned}$$

and the last integral is the Beta function $\beta(t,s) = \dfrac{\Gamma(t)\,\Gamma(s)}{\Gamma(t+s)}$ (see Erdélyi [1953]), where the equality is by the standard formula linking the Beta and Gamma functions. Thus $I^t * I^s = I^{t+s}$ for all $s,t > 0$, and so for all $s,t \in H$ because of the analyticity of $t \mapsto I^t$.

If $0 < t < r = \delta e^{-1} < 1$ and $\delta \le w$, then $w^t \le (w/t)^t \le (w/r)^r$ by considering the derivative of $t \mapsto \log(w/t)$. Thus

$$\begin{aligned}
&\int_{|w|\ge\delta} I^t(w)\,dw \\
&= \int_\delta^\infty \frac{w^{t-1}e^{-w}}{\Gamma(t)}\,dw \\
&\le \frac{r^{-r}}{\Gamma(t)} \int_\delta^\infty w^{r-1}e^{-w}\,dw \\
&\le r^{-r}\,\Gamma(r)\,\Gamma(t)^{-1},
\end{aligned}$$

which tends to zero as t decreases to zero for all $\delta > 0$. Since $I^t(w) > 0$ and since $\|I^t\|_1 = 1$ for all $t > 0$, property (i) follows from Lemma 2.8.

Since the Gelfand map on $L^1(\mathbb{R}^+)$ is essentially the Laplace transform,

$$\sigma(a^t) = \{0\} \cup I^{t\wedge}(\Phi) = \{0\} \cup \{LI^t(z) : z \in H^-\}$$

for all $t \in H$. To prove (iii) it is sufficient to show that $(LI^t)(z) = (z+1)^{-t}$ for all $z \in H^-$ and $t \in H$. For fixed $t \in H$, the functions $z \mapsto LI^t(z)$ and $z \mapsto (z+1)^{-t}$ are analytic in H and continuous in H^-, and from properties of analytic functions we need only show that $LI^t(z) = (z+1)^{-t}$ for all $t \in H$ and $z \ge 0$. Repeating this argument holding $z \ge 0$ fixed and varying t, we are done because

$$\begin{aligned}
LI^t(z) &= \int_0^\infty \frac{w^{t-1}e^{-w}}{\Gamma(t)}\,e^{-wz}\,dw \\
&= \int_0^\infty \frac{u^{t-1}e^{-u}}{\Gamma(t)}\,du\,(z+1)^{-t} \\
&= (z+1)^{-t} \quad \text{for all } t > 0.
\end{aligned}$$

From the asymptotic formula for $\Gamma(x+iy)$ given before Theorem 2.6, we obtain

$$\|I^{x+iy}\|_1 = \Gamma(x).|\Gamma(x+iy)|^{-1}$$

$$= |(x+iy)^{-x-iy}|.\ (2\pi)^{-1/2}.\ \Gamma(x).\ |x+iy|^{1/2}.\ \exp(x + O(|x+iy|^{-1}))$$

$$= (x^2 + y^2)^{-\frac{x}{2} + \frac{1}{4}}\ (2\pi)^{-1/2}.\Gamma(x).\ \exp(y \text{ Arg } (x+iy) + x + O(|x+iy|^{-1}))$$

Since $y \text{ Arg } (x+iy) = |y|\pi/2 - O(1)$ as $|y| \to \infty$, we have

$$\|I^{x+iy}\|_1 = K(x)(1 + y^2/x^2)^{-\frac{x}{2} + \frac{1}{4}}.\ \exp(\pi|y|/2 + O(1))$$

as $|y| \to \infty$ for each $x > 0$. This proves Theorem 2.6.

The following semigroup will be used in the proof of various properties of the Poisson semigroup in Theorem 2.17.

2.9 LEMMA

If $C^t \in L^1(\mathbb{R}^+)$ is defined by

$$C^t(w) = \frac{t}{2\pi^{1/2}} w^{-3/2} \exp(-t^2/4w)$$

for all $w > 0$ and all $t \in Q = \{z \in \mathbb{C} : z \neq 0, |\text{Arg } z| < \pi/4\}$, then $t \mapsto C^t : Q \to L^1(\mathbb{R}^+)$ is an analytic semigroup with the following properties

(i) $(C^t * L^1(\mathbb{R}^+))^- = L^1(\mathbb{R}^+)$ for all $t \in Q$.

(ii) $\|C^t\|_1 = \left(\frac{x^2 + y^2}{x^2 - y^2}\right)^{1/2}$ for $t = x+iy \in Q$.

(iii) $(\mathcal{L}C^t)(z) = e^{-tz^{1/2}}$ for $t \in Q$ and $z \in H^-$.

Before proving this Lemma we evaluate an integral that is closel[y] related to the Laplace transform of C^t.

2.10 LEMMA

If $\alpha > 0$, then

$$e^{-\alpha} = \frac{1}{\pi^{1/2}} \int_0^\infty u^{-1/2} \exp(-u - \alpha^2/4u)\, du$$

$$= \frac{\alpha}{2\pi^{1/2}} \int_0^\infty u^{-3/2} \exp(-u - \alpha^2/4u)\, du.$$

Proof. Let $F(\alpha) = \frac{1}{\pi^{1/2}} \int_0^\infty u^{-1/2} \exp(-u - \alpha^2/4u)du$ for all $\alpha \geq 0$

By the dominated convergence theorem, or uniform convergence,

$$\frac{dF}{d\alpha}(\alpha) = \frac{-\alpha}{2\pi^{1/2}} \int_0^\infty u^{-3/2} \exp(-u - \alpha^2/4u)\, du$$

for all $\alpha > 0$. Let $w = \alpha^2/4u$ in this integral and on simplification we obtain

$$\frac{dF}{d\alpha}(\alpha) = -\frac{1}{\pi^{1/2}} \int_0^\infty w^{-1/2} \exp(-\alpha^2/4w - w)\, dw.$$

Thus F satisfies the differential equation $\frac{dF}{d\alpha} + F = 0$, and so $F(\alpha) = C\, e^{-\alpha}$ for all $\alpha > 0$. The dominated convergence theorem may be used to check the continuity of F at 0, and since $F(0) = \frac{1}{\pi^{1/2}} \Gamma(1/2) = 1$ the proof is complete.

Proof of Lemma 2.9. If $t = x + iy \in Q$ with x and y real, then C^t is a continuous function on $(0,\infty)$, and after making the substitution $\frac{x^2 - y^2}{4w} = u^2$ we obtain

$$\|C^t\|_1 = \frac{|t|}{2\pi^{1/2}} \int_0^\infty w^{-3/2} \exp(-(x^2 - y^2)/4w)\, dw$$

$$= \frac{|t|}{(x^2 - y^2)^{1/2}} \frac{2}{\pi^{1/2}} \int_0^\infty e^{-u^2}\, du$$

$$= \left(\frac{x^2 + y^2}{x^2 - y^2}\right)^{1/2}.$$

Thus $c^t \in L^1(\mathbb{R}^+)$ and $t \mapsto \|c^t\|_1 : Q \to \mathbb{R}$ is a continuous function. Further $t \mapsto c^t(w) : H \to \mathbb{C}$ is analytic and $\frac{\partial c^t}{\partial t}(w) = (1 - t/2w)\, c^t(w)$ for all $w > 0$. Thus $t \mapsto c^t : Q \to L^1(\mathbb{R}^+)$ is an analytic function by Lemma 2.7.

If $t, z > 0$, then the substitution $u = wz$ gives

$$(Lc^t)(z) = \frac{t}{2\pi^{1/2}} \int_0^\infty w^{-3/2} \exp(-wz - t^2/4w)\, dw$$

$$= \frac{tz}{2\pi}^{1/2} \int_0^\infty u^{-3/2} \exp(-u - zt^2/4u)\, du$$

$$= \exp(-z^{1/2}t)$$

by Lemma 2.10. The analyticity of the functions $(Lc^t)(z)$ and $\exp(-z^{1/2}t)$ implies that they are equal for $t \in Q$ and $z \in H^-$. The semigroup property follows from the one-to-one property of the Laplace transform and $\exp(-(t+s)z^{1/2}) = \exp(-tz^{1/2}).\exp(-sz^{1/2})$ for all $z \in H^-$. If $t, \delta > 0$, then

$$\int_{|w| \geq \delta} c^t(w)\, dw \leq \frac{t}{2\pi^{1/2}} \int_{w \geq \delta} w^{-3/2}\, dw$$

$$= \frac{t}{2(\pi\delta)^{1/2}},$$

which tends to zero as t decreases to zero for fixed $\delta > 0$. Thus property (i) holds by Lemma 2.8. We have already checked (ii) and (iii).

2.11 REMARKS, AND OTHER SEMIGROUPS IN $L^1(\mathbb{R}^+)$

The order of growth of $\|I^{x+iy}\|_1$ for fixed $x > 0$ as $|y| \to \infty$ is the same as $\|a^{x+iy}\|$ in Property 6 of Theorem 3.1. It is possible that this is essentially the best order of growth along a vertical line for an analytic semigroup in a Banach algebra in general. Special algebras like C^*-algebras (Theorem 2.3) or $L^1(\mathbb{R}^n)$ have semigroups with considerably better orders of growth. Note that each non-zero analytic semigroup $t \mapsto a^t : H \to L^1(\mathbb{R}^+)$ has

$$\int_{\mathbb{R}} (1 + y^2)^{-1} \log^+ \|a^{1+iy}\|_1\, dy$$

divergent by Theorem 5.6 because there are continuous monomorphisms from $L^1(\mathbb{R}^+)$ into radical Banach algebras (see 2.12 and 3.6).

We shall mention some other semigroups into $L^1(\mathbb{R}^+)$ without details. Given a function $t \mapsto f^t$ from some additive subgroup of H into $L^1(\mathbb{R}^+)$ to prove the semigroup property $f^{s+t} = f^s * f^t$ it is sufficient to show that $(\mathcal{L}f^t)(z) = \exp tG(z)$ for some analytic function G on the open right half plane H. A study of the Laplace transforms in the Bateman Project table of integrals (Erdélyi [1954]) shows that the following tables give semigroups from H into $L^1(\mathbb{R}^+)$:

#20, p.238 with $t = \lambda = \nu \in H$

#13, p.239 with $t = -\nu \in H$

#15, p.239 with $t = \nu \in H$

#21, p.240 with $t = \nu \in H$.

2.12 THE RADICAL CONVOLUTION ALGEBRAS $L^1(\mathbb{R}^+,\omega)$

Let ω be a continuous function from $\mathbb{R}^+$ into $(0,\infty)$ satisfying $\omega(0) = 1$, $\omega(s+t) \le \omega(s).\omega(t)$ for all s and $t \in \mathbb{R}^+$, and $\omega(t)^{1/t} \to 0$ as $t \to \infty$. Such a function is called a radical weight. For example, $\omega(t) = e^{-t^2}$, $\omega(t) = e^{-t \log t}$, and $\omega(t) = e^{-t \log \log t}$ for large t are radical weight functions. Let $L^1(\mathbb{R}^+,\omega)$ be the Banach space of equivalence classes of locally integrable functions f on $\mathbb{R}^+$ such that $\|f\| = \int_{\mathbb{R}^+} |f(w)| \ \omega(w)\, dw$ is finite. With the convolution product $f * g(t) = \int_0^t f(t-w)\, g(w)\, dw$ for almost all $t \in \mathbb{R}^+$, this Banach space becomes a commutative radical Banach algebra. The Titchmarsh convolution theorem may be used to show that $L^1(\mathbb{R}^+,\omega)$ is an integral domain. Further the Banach algebra $L^1(\mathbb{R}^+,\omega)$ has a countable bounded approximate identity bounded by 1 (because $\omega(0) = 1$), and the identity map from $L^1(\mathbb{R}^+)$ into $L^1(\mathbb{R}^+,\omega)$ is a continuous monomorphism from $L^1(\mathbb{R}^+)$ onto a dense subalgebra of $L^1(\mathbb{R}^+,\omega)$. This monomorphism has norm 1 if and only if $\omega(t) \le 1$ for all $t > 0$. The analytic semigroups in $L^1(\mathbb{R}^+)$ are mapped into analytic semigroups in $L^1(\mathbb{R}^+,\omega)$ by the identity map; however the spectra and some of the norm properties change in the process.

2.13 THE VOLTERRA ALGEBRA $L^1_*[0,1]$

The Banach space $L^1[0,1]$ becomes a commutative radical Banach algebra with the product $(f*g)(t) = \int_0^t f(t-w)\, g(w)\, dw$ for almost all $t \in [0,1]$. We denote this Banach algebra by $L^1_*[0,1]$, and call it the Volterra algebra. This Banach algebra has a countable bounded approximate identity bounded by 1, and the restriction map $f \mapsto f|_{[0,1]}$: $L^1(\mathbb{R}^+) \to L^1_*[0,1]$ is a norm reducing algebra epimorphism. In fact the restriction map is a continuous algebra epimorphism from each radical algebra $L^1(\mathbb{R}^+,\omega)$ of 2.12 onto $L^1_*[0,1]$. The analytic semigroups in $L^1(\mathbb{R}^+)$ are carried into analytic semigroups in $L^1_*[0,1]$ by the restriction map. Note that the Volterra algebra has a dense ideal of nilpotent elements : the set $\{f \in L^1_*[0,1] : f = 0$ a.e. on $[0,\varepsilon]$ for some $\varepsilon > 0\}$.

2.14 THE CONVOLUTION ALGEBRA $L^1(\mathbb{R}^n)$

We shall now discuss two very important semigroups, the Gaussi (or Weierstrass) and Poisson (or Cauchy) semigroups, in the convolution Banach algebra $L^1(\mathbb{R}^n)$, where n is a positive integer. These two semigroups were the motivation for several of the properties of Theorem 3.1, and for the Tauberian Theorem 5.6. They play an important role in probability theory and in several branches of analysis but we shall restri our attention here to properties related to the Banach algebras $L^1(\mathbb{R}^n)$. W shall not discuss the corresponding semigroups in $L^1(G)$ for G a Lie group; for an excellent account of this topic see Chapter 2 of Stein [1970

We shall recall some well known facts about $L^1(\mathbb{R}^n)$ and at th same time introduce notation. The Banach algebra $L^1(\mathbb{R}^n)$ consists of (equivalence classes of) integrable functions f on $\mathbb{R}^n$ with norm $\|f\|_1 = \int_{\mathbb{R}^n} |f(w)|dw$, and convolution product $f * g(w) = \int_{\mathbb{R}^n} f(w-u)\, g(u)$ for almost every $w \in \mathbb{R}^n$. The algebra $L^1(\mathbb{R}^n)$ is a commutative Banach algebra without identity and with a countable bounded approximate identity bounded by 1. There is a natural involution $*$ on $L^1(\mathbb{R}^n)$ defined by $f^*(w) = \overline{f(-w)}$ for all $f \in L^1(\mathbb{R}^n)$, and with this involution $L^1(\mathbb{R}^n)$ is Banach $*$-algebra. The algebra $L^1(\mathbb{R}^n)$ is (Jacobson) semisimple, and its carrier space may be identified with $\mathbb{R}^n$ by the mapping $\lambda \mapsto \phi_\lambda : \mathbb{R}^n \to \Phi$ defined by $\phi_\lambda(f) = \int_{\mathbb{R}^n} f(w) \exp(-2\pi i\langle\lambda,w\rangle)\, dw$ for all $f \in L^1(\mathbb{R}^n)$ whe $\langle\lambda,w\rangle$ is the usual inner product of λ and w in $\mathbb{R}^n$. With this identification the Gelfand transform from $L^1(\mathbb{R}^n)$ into $C_0(\Phi)$ becomes t Fourier transform

$$f^{\wedge}(\phi_\lambda) = \phi_\lambda(f) = \int_{\mathbb{R}^n} f(w) \exp(-2\pi i\langle\lambda,w\rangle)\, dw = f^{\wedge}(\lambda).$$

The inverse Fourier transform is $f^{\vee}(\lambda) = f^{\wedge}(-\lambda)$ for all $f \in L^1(\mathbb{R}^n)$ and $\lambda \in \mathbb{R}^n$. The multiplier algebra $\mathrm{Mul}(L^1(\mathbb{R}^n))$ of $L^1(\mathbb{R}^n)$ is naturally isometrically isomorphic with the convolution measure algebra $M(\mathbb{R}^n)$, which is defined in a similar way to $M(\mathbb{R}^+)$. The Laplacian Δ is the differential operator $\sum_1^n \frac{\partial^2}{\partial x_j^2}$ defined on suitably differentiable functions on $\mathbb{R}^n$.

2.15 <u>THEOREM</u>

If $G^t(w)$ is defined by $G^t(w) = (4\pi t)^{-n/2} e^{-|w|^2/4t}$ for all $w \in \mathbb{R}^n$ and all $t \in H$, then $t \mapsto G^t : H \to L^1(\mathbb{R}^n)$ is an analytic semigroup, the Gaussian (or Weierstrass) semigroup, with the following properties.

(i) $(G^t * L^1(\mathbb{R}^n))^- = L^1(\mathbb{R}^n)$ for all $t \in H$.

(ii) $\|G^{x+iy}\|_1 = (1 + y^2/x^2)^{n/4}$ for all $x + iy \in H$.

(iii) $G^{t\wedge}(\lambda) = \exp(-4\pi^2|\lambda|^2 t)$ for all $\lambda \in \mathbb{R}^n$, and $\sigma(G^t) = \{0\} \cup \{\exp(-t\alpha) : \alpha \geq 0\}$ for all $t \in H$.

(iv) $G^t \geq 0$ as a function and as an element of the *-algebra $L^1(\mathbb{R}^n)$ for all $t > 0$.

(v) $(\frac{\partial}{\partial t} - \Delta)(G^t * f) = 0$ for all $f \in L^1(\mathbb{R}^n)$.

<u>Proof</u>. For notational convenience we shall prove this theorem for $n = 1$ only. Minor changes give the general case, and when we discuss the Poisson semigroup in Theorem 2.17 we shall consider the case n a positive integer. In this proof we shall omit the range of integration if it is $\mathbb{R}$. A trivial substitution turns $\int \exp(-w^2)dw = \pi^{1/2}$ into $\int \exp(-\alpha w^2)dw = (\pi/\alpha)^{1/2}$ for each $\alpha > 0$.

If $t = x + iy \in H$, then G^t is a continuous function on $\mathbb{R}$, and

$$\begin{aligned}\|G^t\|_1 &= (4\pi|t|)^{-1/2} \int e^{-w^2x/4|t|^2}\, dw \\ &= (4\pi|t|)^{-1/2} (4\pi|t|^2/x)^{1/2} \\ &= (1 + y^2/x^2)^{1/4}.\end{aligned}$$

Thus $G^t \in L^1(\mathbb{R})$ for all $t \in H$, and $t \mapsto \|G^t\|_1 : H \to \mathbb{R}$ is continuous.

Further $t \mapsto G^t(w) : H \to \mathbb{C}$ is analytic and $\dfrac{\partial G^t(w)}{\partial t} = G^t(w) \left\{ \dfrac{w^2}{4t^2} - \dfrac{1}{2t} \right\}$ for each $w \in \mathbb{R}$. By Lemma 2.7, $t \mapsto G^t : H \to L^1(\mathbb{R})$ is an analytic function.

If $t, r > 0$ and $w, u \in \mathbb{R}$, then we complete the square

$$\{(w-u)^2 t^{-1} + u^2 r^{-1}\} = \frac{t+r}{tr}\left(u - \frac{rw}{t+r}\right)^2 + \frac{w^2}{t+r} ,$$

and using this we obtain

$$\begin{aligned}
& G^t * G^r(w) \\
&= (4\pi)^{-1} (tr)^{-1/2} \int \exp\{-(w-u)^2 4^{-1} t^{-1} - u^2 4^{-1} r^{-1}\}\, du \\
&= (4\pi)^{-1} (tr)^{-1/2} \exp(-w^2/4(t+r)) \\
&\qquad \int \exp\{-(t+r) t^{-1} r^{-1} 4^{-1} v^2\}\, dv , \\
&\text{where } v = u - \frac{rw}{t+r} , \\
&= (4\pi)^{-1} (tr)^{-1/2} \exp(-w^2/4(t+r)) \pi^{1/2} (tr/(t+r))^{1/2} \\
&= G^{t+r}(w).
\end{aligned}$$

Since $t \mapsto G^t : H \to L^1(\mathbb{R})$ is an analytic function, the semigroup property extends from $(0,\infty)$ to H. Alternatively the semigroup property may be checked using the Fourier transform of G^t and the semisimplicity of $L^1(\mathbb{R})$

If $t, \delta > 0$, then $\exp(w^2/4t) \geq w^2/4t$ and so

$$\int_{|w| \geq \delta} G^t(w)\, dw \leq \frac{4t}{(4\pi t)^{1/2}} \int_{|w| \geq \delta} \frac{1}{w^2}\, dw = 4t^{1/2}/\pi^{1/2}\delta ,$$

which tends to zero as t tends to zero for each $\delta > 0$. By Lemma 2.8 it follows that $(G^t * L^1(\mathbb{R}))^- = L^1(\mathbb{R})$ for all $t \in H$.

Let $F(t,z) = (4\pi t)^{-1/2} \int \exp(-2\pi z w - w^2 4^{-1} t^{-1})\, dw$ for all $z \in \mathbb{C}$ and $t > 0$. Note that the integral converges because of the rapid decay of $w \mapsto \exp(-w^2 4^{-1} t^{-1})$ near infinity, and that $z \mapsto F(t,z)$ is an entire function for all $t \in H$. If z is in $\mathbb{R}$, then

$$\begin{aligned}
F(t,z) &= (4\pi t)^{-1/2} \int \exp(-(w+4\pi t z)^2 4^{-1} t^{-1})\, dw \exp(4\pi^2 z^2 t) \\
&= \exp(4\pi^2 z^2 t).
\end{aligned}$$

Thus $F(t,z) = \exp(4\pi^2 z^2 t)$ for all $z \in \mathbb{C}$, and hence $G^{t\wedge}(\lambda) = F(t,i\lambda)$ $= \exp(-4\pi^2\lambda^2 t)$ for all $\lambda \in \mathbb{R}$ and $t > 0$. Using analyticity again this formula holds for all $t \in H$.

From the definition of G^t it is clear that $G^t \geq 0$ as a function for all $t > 0$. Also $(G^t)^* = G^t$ so that $G^t = (G^{t/2})^* * G^{t/2} \geq 0$ for all $t > 0$. Part (v) follows from the definition of the convolution and the formula

$$\frac{\partial G^t}{\partial t}(w - u) = G^t(w - u)\left\{\frac{(w - u)^2}{4t^2} - \frac{1}{2t}\right\}$$

$$= \frac{\partial^2 G}{\partial w^2}(w - u).$$

This completes the proof.

Before we discuss the Poisson semigroup we give a standard little lemma for evaluating spherically symmetric integrals over $\mathbb{R}^n$. If we had proved Theorem 2.15 on the Gaussian semigroup for $n > 1$, this lemma would have been useful.

2.16 <u>LEMMA</u>

The area of the surface of the closed unit sphere in $\mathbb{R}^n$ is $\omega_n = 2\pi^{n/2}\,\Gamma(n/2)^{-1}$. If the function $r \mapsto f(r)r^{n-1} : (0,\infty) \to \mathbb{C}$ is in $L^1(\mathbb{R}^+)$, then $w \mapsto f(|w|) : \mathbb{R}^n \to \mathbb{C}$ is in $L^1(\mathbb{R}^n)$ and

$$\int_{|w|\leq k} f(|w|)\, dw = \omega_n \int_0^k f(r)r^{n-1}\, dr$$

for $0 < k \leq \infty$

<u>Proof</u>. Let $V(r)$ denote the volume of the sphere of radius r in $\mathbb{R}^n$. The surface area of the sphere of radius r in $\mathbb{R}^n$ is $r^{n-1}\,\omega_n$. From the definition of the derivative and the volume of a thin shell being approximately the area of the shell times its thickness, we have $\frac{dV}{dr} = r^{n-1}\,\omega_n$. Since $w \mapsto f(|w|) : \mathbb{R}^n \to \mathbb{C}$ is constant on spheres with centre the origin, $\int_{|w|\leq k} f(|w|)\, dw = \omega_n \int_0^k f(r)\, r^{n-1}\, dr$

from the definition of the integral. Alternatively we can write each $w \in \mathbb{R}^n \backslash \{0\}$ uniquely in the form $w = rs$, where $0 < r < \infty$ and $s \in S = \{u \in \mathbb{R}^n : |u| = 1\}$, so $dw = r^{n-1}\, dr\, d\sigma(s)$, where $d\sigma$ is the area measur on S. The equality of the integrals then follows from this.

We calculate ω_n by applying the integral equality with $f(r) = e^{-\pi r^2}$ for $r \geq 0$. Letting $\pi r^2 = u$ in the r-integral we obtain

$$\int_{\mathbb{R}^n} e^{-\pi|w|^2}\, dw = \frac{\omega_n}{2\pi^{n/2}} \int_0^\infty e^{-u} u^{n/2-1}\, du$$

$$= \frac{\omega_n \Gamma(n/2)}{2\pi^{n/2}} \quad .$$

If $w = (w_1, \ldots, w_n) \in \mathbb{R}^n$, then

$$\int_{\mathbb{R}^n} e^{-\pi|w|^2}\, dw$$

$$= \prod_{j=1}^{n} \int_{\mathbb{R}} e^{-\pi w_j^2}\, dw_j$$

$$= \left\{ \int_{\mathbb{R}^2} e^{-\pi(w_1^2 + w_2^2)}\, d(w_1, w_2) \right\}^{n/2}$$

$$= \left\{ \frac{\omega_2}{2\pi} \Gamma(2/2) \right\}^{n/2}$$

by our formula above with $n = 2$. Since ω_2 is the length of the circumference of the unit circle, $\int_{\mathbb{R}^n} e^{-\pi|w|^2}\, dw = 1$. This completes the proof of the Lemma.

Observe that ω_n is always a rational multiple of an integer power of π because $\Gamma(n/2)$ is a rational multiple of $\pi^{1/2}$ if n is odd, and is an integer if n is even.

2.17 <u>THEOREM</u>

If $P^t(w)$ is defined by

$$P^t(w) = \frac{\Gamma((n+1)/2)}{\pi^{(n+1)/2}} \cdot \frac{t}{(t^2+|w|^2)^{(n+1)/2}}$$

for all $w \in \mathbb{R}^n$ and all $t \in H$, then $t \mapsto P^t : H \to L^1(\mathbb{R}^n)$ is an analytic semigroup, called the Poisson (or Cauchy) semigroup, with the following properties.

(i) $(P^t * L^1(\mathbb{R}^n))^- = L^1(\mathbb{R}^n)$ for all $t \in H$.

(ii) $\|P^t\|_1 = 1$ for all $t > 0$,

$\{|y|^{\frac{1-n}{2}} \|P^{1+iy}\|_1 : y \in \mathbb{R}, |y| \geq 1\}$ is bounded for each $n \geq 2$, and $\{(\log|y|)^{-1} \|P^{1+iy}\|_1 : y \in \mathbb{R}, |y| \geq 1\}$ is bounded for $n = 1$.

(iii) $P^{t\wedge}(\lambda) = \exp(-2\pi|\lambda|t)$ and $\sigma(P^t) = \{0\} \cup \{\exp(-rt) : r > 0\}$ for all $t \in H$ and $\lambda \in \mathbb{R}^n$.

(iv) $P^t \geq 0$ as a function and as an element of the *-algebra $L^1(\mathbb{R}^n)$ for all $t > 0$.

(v) $P^t = \int_0^\infty C^t(u)\, G^u\, du$ for all $t \in Q = \{z \in H : |\mathrm{Arg}\, z| < \pi/4\}$.

<u>Proof</u>. If $t \in H$, then the function $r \mapsto (t^2+r^2)^{-(n+1)/2} : \mathbb{R}^+ \to \mathbb{C}$ is continuous and $|t^2+r^2|^{-(n+1)/2} \cdot r^{n-1} = O(r^{-2})$ as r tends to infinity. By Lemma 2.16 it follows that $P^t \in L^1(\mathbb{R}^n)$ and that

$$\|P^t\|_1 = \frac{2\Gamma((n+1)/2)}{\pi^{1/2}\Gamma(n/2)} \int_0^\infty \frac{|t|\, r^{n-1}}{|t^2+r^2|^{(n+1)/2}}\, dr. \tag{1}$$

From (1) we see that $t \mapsto \|P^t\|_1 : H \to \mathbb{R}$ is a continuous function. Also for each $w \in \mathbb{R}^n$ the function $t \mapsto P^t(w) : H \to \mathbb{C}$ is analytic, and $(t,w) \mapsto P^t(w) : H \times \mathbb{R}^n \to \mathbb{C}$ is continuous. By Lemma 2.7 $t \mapsto P^t : H \to L^1(\mathbb{R}^n)$ is an analytic function.

We shall now check property (v) from which the semigroup property and the Fourier transform of P^t will follow easily. If $t > 0$, the substitution $\zeta = (t^2+|w|^2)/4u$ reduces

$$\int_0^\infty C^t(u)\, G^u(w)\, du$$

$$= 2^{-1}\, \pi^{-1/2} \int_0^\infty t\, u^{-3/2}\, e^{-t^2/4u}\, (4\pi u)^{-n/2}\, e^{-|w|^2/4u}\, du$$

to

$$t\ \pi^{-(n+1)/2}(t^2+|w|^2)^{-(n+1)/2}\int_0^\infty \zeta^{(n-1)/2}\ e^{-\zeta}d\zeta = P^t(w)$$

by definition of the Gamma function and Poisson semigroup. Since $C^t \in L^1(\mathbb{R}^+)$, the integral $\int_0^\infty C^t(u)\ G^u\ du$ exists as a Bochner integral in $L^1(\mathbb{R}^n)$ and is equal to P^t by the above. The analyticity of $t \mapsto C^t : Q \to L^1(\mathbb{R}^+)$ and the continuity of the operator

$\theta : f \mapsto \int_0^\infty f(u)\ G^u\ du : L^1(\mathbb{R}^+) \to L^1(\mathbb{R}^n)$ imply that the function

$t \mapsto \int_0^\infty C^t(u)\ G^u\ du : Q \to L^1(\mathbb{R}^n)$ is analytic. Property (v) follows from this and the analyticity of $t \mapsto P^t$. The operator θ from $L^1(\mathbb{R}^+)$ into $L^1(\mathbb{R}^n)$ may be seen to be a homomorphism using Fubini's Theorem. Since $P^t = \theta(C^t)$ for all $t \in Q$, we see that $t \mapsto P^t$ is a semigroup on Q and hence on H.

The continuity of the Fourier transform from $L^1(\mathbb{R}^n)$ into $C_o(\mathbb{R}^n)$ (or Fubini's Theorem) shows that

$$\begin{aligned} P^{t\wedge}(\lambda) &= \int_0^\infty C^t(u)\ G^{u\wedge}(\lambda)\ du \\ &= \int_0^\infty C^t(u)\ \exp(-4\pi^2|\lambda|^2 u)\ du \\ &= (LC^t)(4\pi^2|\lambda|^2) \\ &= \exp(-2\pi|\lambda|t) \end{aligned}$$

for $t \in Q$ and $\lambda \in \mathbb{R}^n$
by Theorem 2.15 and Lemma 2.9. Using analyticity $P^{t\wedge}(\lambda) = \exp(-2\pi|\lambda|t)$ for all $\lambda \in \mathbb{R}^n$ and $t \in H$. By Fubini's Theorem we have

$$\|P^t\|_1 \le \int_0^\infty |C^t(u)|.\|G^u\|_1\ du \le \|C^t\|_1$$

for all $t \in Q$ since $\|G^u\|_1 = 1$ for $u > 0$. Hence $\|P^t\|_1 = 1$ for all $t > 0$ because $\|P^t\|_1 \ge P^{t\wedge}(0) = 1$. Since $C^t(u) \ge 0$ for all $t, u > 0$, we see that $P^t \ge 0$ for $t > 0$ follows from the corresponding property of G^t.

To prove property (i) it is sufficient to show that for each $\delta > 0$, $\int_{|w|\geq\delta} P^t(w)\,dw \to 0$ as $t \to 0$, $t > 0$ (Lemma 2.8). Discarding the constants in P^t and using Lemma 2.16 we see that it is sufficient to prove that

$$\int_\delta^\infty t(t^2+r^2)^{-(n+1)/2}\, r^{n-1}\, dr \to 0$$

as $t \to 0$, $t > 0$, for each $\delta > 0$. For $0 < t < \delta$, we have

$$\int_\delta^\infty t(t^2+r^2)^{-(n+1)/2}\, r^{n-1} dr \leq \int_\delta^\infty t2^{-(n+1)/2}.r^{-(n+1)}\, r^{n-1} dr$$
$$\to 0 \quad \text{as} \quad t \to 0.$$

We shall now obtain an upper bound for $\|P^{1+iy}\|_1$ for $y \geq 4$ - this will complete the proof of (ii). From (1) $\|P^{1+iy}\|_1$ is a constant times

$$I = \int_0^\infty \frac{|1+iy|\; r^{n-1}}{|(1+iy)^2 + r^2|^{(n+1)/2}}\, dr$$

and we estimate this integral. Since $|y| > 1$ we have

$$2|(1+iy)^2+r^2| \geq \begin{cases} (y^2-1-r^2+2y) & \text{if } r \leq (y^2-1)^{1/2} \\ (r^2-y^2+1+2y) & \text{if } r \geq (y^2-1)^{1/2} \end{cases}$$

Using these inequalities and replacing $|1 + iy|$ by $2y$, we have $I \leq 2^{(n+1)1/2}\,(I_1 + I_2)$ where

$$I_1 = 2y \int_0^{(y^2-1)^{1/2}} \frac{r^{n-1}}{(y^2-1-r^2+2y)^{(n+1)/2}}\, dr$$

and

$$I_2 = 2y \int_{(y^2-1)^{1/2}}^\infty \frac{r^{n-1}}{(r^2-y^2+1+2y)^{(n+1)/2}}\, dr\,.$$

The substitution $r = (y^2+2y-1)^{1/2}\cos\xi$ reduces I_1 to

$$2y\ (y^2+2y-1)^{-1/2}\int_{\nu}^{\pi/2} \sin^{-n}\xi\ d\xi,$$

where $\sin\nu = (2y/(y^2-1))^{1/2}$. In the second integral we use the substitution $r = (y^2-2y-1)^{1/2}\sec\zeta$ for $y > 3$ to reduce I_2 to

$$2y(y^2-2y-1)^{-1/2}\int_{\nu}^{\pi/2} \sin^{-n}\zeta\ d\zeta.$$

From the graph of the sine function we obtain the inequality $\frac{2\xi}{\pi} \leq \sin\xi$ for $0 \leq \xi \leq \pi/2$. Using this inequality $\int_{\nu}^{\pi/2} \sin^{-n}\xi\ d\xi \leq (\pi/2)^n\ \nu^{-n+1}\ (n-1)$ if $n \geq 2$ (or $(\pi/2)\log(\nu^{-1})$ if $n = 1$) neglecting the term arising from $\pi/2$. We now use $(2y/(y^2-1))^{1/2} = \sin\nu \geq 2\pi^{-1}\nu$ and $y^2+2y-1 \geq y^2-2y-1 \geq y^2/4$ for $y \geq 4$, to deduce that

$$I \leq 2^{(n+1)/2}.2.2y(y^2/4)^{-1/2}(\pi/2)^n(n-1)^{-1}(\pi/2)^{-n+1}((y^2-1)/2y)^{(n-1}$$
$$\leq 8\pi(n-1)^{-1}\ y^{(n-1)/2}$$

for all $y \geq 4$ and $n \geq 2$ (and $I \leq 4\pi\log y$ when $n = 1$). This proves (ii) since $\|P^{1+iy}\|_1 = \|P^{1-iy}\|_1$ for all $y > 0$.

2.18 <u>REMARKS</u>

Compare the rates of growth $\|P^{1+iy}\|_1 = O(|y|^{\frac{n-1}{2}})$ and $\|G^{1+iy}\|_1 = O(|y|^{n/2})$ for $n \geq 2$ for the Poisson and Gaussian semigroups as $|y|$ tends to infinity (for $n = 1$ the Poisson growth is $\log|y|$). Intuitively the subordination of P^t to G^t has smoothed out G^t so giving a lower rate of growth for P^t.

There is a natural isometric embedding ψ from $L^1(\mathbb{R}^+)$ into $L^1(\mathbb{R})$ given by $(\psi f)(x) = f(x)$ if $x \geq 0$ and $(\psi f)(x) = 0$ if $x < 0$, and we shall regard $L^1(\mathbb{R}^+)$ as a closed subalgebra of $L^1(\mathbb{R})$ via this embedding. Since $(L^1(\mathbb{R}^+) * L^1(\mathbb{R}))^- = L^1(\mathbb{R})$ (by Cohen's factorization theorem - see the introduction), we see that the fractional integral semigroup $t \to I^t : H \to L^1(\mathbb{R})$ has many of the properties of the Gaussian and Poisson semigroups. However $(I^t)^* \neq I^t$ for $t > 0$, because $L^1(\mathbb{R}^+) \cap L^1(\mathbb{R})^+)^* = \{0\}$, and the order of growth of $\|I^{1+iy}\|$, namely

$O(y^{-1/2} \exp(\Pi|y|/2))$, as $|y|$ tends to infinity is substantially worse than the polynomial growth of the Gaussian or Poisson semigroups. Is there a natural isometric algebraic embedding of $L^1(\mathbb{R}^+)$ into $L^1(\mathbb{R}^n)$ for $n > 1$ not of the type discussed in Corollary 3.5?

2.19 NOTES AND REMARKS

Many of the semigroups discussed in this chapter are considered in depth by Hille and Phillips [1974]. However we do not discuss generators and we emphasize different aspects of the theory than Hille and Phillips [1974], where there are further references and remarks.

C^*-algebras. Aarnes and Kadison [1969] prove that a separable C^*-algebra has a commutative bounded approximate identity. There is a discussion of commutative approximate identities in C^*-algebras in Pedersen [1979] (see Corollary 3.12.15) that essentially gives Theorem 2.2. See also Doran and Wichman [1979]. See also Lemma 4.3 of Haagerup [1981].

The convolution algebra $L^1(\mathbb{R}^+)$. The fractional integral semigroup $t \mapsto I^t$ has played a fundamental role in harmonic analysis - see Hardy and Littlewood [1932] and Stein [1970]. This semigroup is used in Stein's treatment of the general maximal function (see p.77 and p.117 in Stein [1970]). When acting on $L^p(\mathbb{R}^+)$ by convolution this semigroup is also called the Riemann-Liouville operator. The semigroup $C^t(w) = t\, w^{-3/2}\, (2\pi^{1/2})^{-1}\, e^{-t^2/4w}$ occurs in the backwards heat equation and there is a discussion of it in Widder [1975]. This semigroup is also used in the subordination of the Poisson semigroup to the heat (Gaussian) semigroup in the group algebra of a Lie group (see Stein [1970], p.46). Esterle [1980e] uses it in the construction of infinitely differentiable semigroups in a radical Banach algebra that tend to zero very fast as t tends to infinity (see 5.4).

The convolution radical algebras $L^1(\mathbb{R}^+,\omega)$ and $L^1{}_*[0,1]$. These Banach algebras have been extensively studied since the early developments in Banach algebra theory (see Gelfand, Raikov, and Shilov [1964], and Hille and Phillips [1974]). Recently there has been renewed interest in these algebras' closed ideals (Allan [1979]), derivations (Ghahramani [1980]), and rates of growth (Allan [1979], Bade and Dales [1980], and Esterle [1980e]).

<u>The convolution algebras</u> $L^1(\mathbb{R}^n)$. The Gaussian and Poisson semigroups play an important role in harmonic analysis and function theory on $\mathbb{R}^n$ as may be seen by examining the books Stein and Weiss [1971] and Stein [1970]. The Gaussian (or normal) semigroup plays an important role in probability theory and the Poisson semigroup in the theory of harmonic functions. However these are outside the scope of these notes and remarks. Computationally the Gaussian semigroup is easier to handle than the Poisson semigroup. For an excellent account of semigroups corresponding to the Gaussian and Poisson semigroups in the group algebra of a Lie group see Stein [1970] (also Hulanicki [1974]).

<u>Lemma 2,7</u>. H.G. Dales has proved the following Lemma, which is simpler to prove and stronger than Lemma 2.7.

<u>Lemma</u>. Let $(W,\sum,\mu)$ be a measure space with μ a positive measure and let $1 \leq p < \infty$. Let $w \mapsto F(t,w)$ be in $L^p(W)$ for each $t \in H$, and let $t \mapsto F(t,w) : H \to \mathbb{C}$ be analytic for each $w \in W$. If for each compact set K contained in H there is a function $\rho_k \in L^p(W)$ so that $|F(t,w)| \leq \rho_k(w)$ for all $t \in K$ and $w \in W$, then $t \mapsto F(t,\cdot) : H \to L^p(W)$ is analytic.

I have used Lemma 2.7 instead of this Lemma because the hypotheses of Lemma 2.7 seem more natural and are easier to check. This Lemma is proved by using Lemma 1.3, the Cauchy estimates, and the dominated convergence theorem.

3 EXISTENCE OF ANALYTIC SEMIGROUPS - AN EXTENSION OF COHEN'S FACTORIZATION METHOD

Throughout this chapter A will denote a Banach algebra with a countable bounded approximate identity, which will usually be bounded by 1. The main theorem of this chapter ensures that such an algebra contains a semigroup analytic in the open right half plane. This theorem is proved by an extension of Cohen's factorization theorem for a Banach algebra with a bounded approximate identity. In this chapter we shall discuss the properties of the semigroups that may be obtained by these methods, and shall investigate some applications of the existence of the semigroups. The proof of the existence of the semigroups is given in detail in Chapter 4, and is discussed there. The basic properties of the semigroup are stated in Theorem 3.1, and additional properties relating to derivations, multipliers and automorphisms are dealt with in Theorem 3.15.

Many of the properties of the semigroups given below are generalizations of those of the fractional integral semigroup in $L^1(\mathbb{R}^+)$, or the Gaussian and Poisson semigroups in $L^1(\mathbb{R}^n)$. We should like the semigroups constructed in A to be as nice as possible: to have good growth, norm and spectral behaviour. In Chapter 5 it is shown that certain polynomial growth properties of the Gaussian and Poisson semigroups cannot hold in radical Banach algebras.

The semigroup properties of a^t are emphasised rather than the factorization properties, and all the semigroup properties that I know are included in Theorems 3.1 and 3.15. However in any given situation one is usually interested in only two or three properties of the semigroup, or of the module factor of x. Most parts of Theorem 3.1 have been used in an application or a stronger version has been used as the hypothesis of a result restricting the structure of the algebra (see Chapter 5).

3.1 <u>THEOREM</u>

Let A be a Banach algebra with a countable bounded approximate identity bounded by 1, and without an identity. Let X (and Y) be a left (and a right) Banach A-module, and let x (and y) be in the closed linea span of $A.X = \{a.w : a \in A, w \in X\}$ (and of $Y.A$). Then an analytic semi group $t \mapsto a^t : H \to A$ from the open right half plane H into A and entire functions $t \mapsto x_t : \mathbb{C} \to X$ (and $t \mapsto y_t : \mathbb{C} \to Y$) may be chosen to satisfy the following properties.

1. $x = a^t.x_t$ (and $y = y_t.a^t$) for all $t \in H$.

Properties of the semigroup

2. $(a^t.A)^- = A = (A.a^t)^-$ for all $t \in H$.
3. For each $\psi \in (0,\pi/2)$ the function $t \mapsto \|a^t\|$ is bounded in the sector $U(\psi) = \{z \in \mathbb{C} : 0 < |z| \leq 1, |\text{Arg } z| \leq \psi\}$, where Arg is the principal argument.
4. For each $\psi \in (0,\pi/2)$ and each $b \in A$, $a^t b \to b$ and $ba^t \to b$ as $t \to 0$ in $U(\psi)$.
5. $\|a^t\| \leq 1$ for all $t > 0$.
6. If $C_\alpha = 1 + 3(1 - 4^{-\alpha})^{-1}$ for $\alpha > 0$, then
$$\|a^{x+iy}\| \leq C_\alpha . 2^x \left(1 + \frac{y^2}{(x-\alpha)^2}\right)^{-(x-\alpha)/2} e^{\pi|y|/2}$$
for all $x > \alpha$ and all $y \in \mathbb{R}$.
7. $\sigma(a^x) \subseteq U(\pi x/2)^- = \{0\} \cup U(\pi x/2)$ for $0 < x < 1$, and $\nu(a^t) \leq \exp(\pi|\text{Im } t|/2)$ for all $t \in H$, where ν is the spectral radius.
8. If $\theta : f \mapsto \int_0^\infty f(t)a^t dt : L^1(\mathbb{R}^+) \to A$, then θ is one-to-one.
9. If A has a closed convex bounded approximate identity Λ bounded by 1 such that $b \in \Lambda$ implies that $b^n \in \Lambda$ for all positive integers n, then $a^t \in \Lambda$ for all $t > 0$.

Properties of the factors in the module

10. $x_0 = x$ and $x_{-t} = a^t.x$ (and, $y_0 = y$ and $y_{-t} = y.a^t$) for all $t \in H$.
11. $a^t.x_{z+t} = x_z$ (and $y_{z+t}.a^t = y_z$) for all $t \in H$ and $z \in \mathbb{C}$.
12. $x_t \in (A.x)^-$ (and $y_t \in (y.A)^-$) for all $t \in \mathbb{C}$.
13. If $\delta > 0$ and if $t \mapsto \alpha_t : (0,\infty) \to [1+\delta,\infty)$ such that $\alpha_t \to \infty$ as $t \to \infty$, then $\|x_t\| \leq (\alpha_{|t|})^{|t|^t} \|x\|$ (and $\|y_t\| \leq (\alpha_{|t|})^{|t|^t} \|y\|$) for all $t \in \mathbb{C}$ with $|t| \geq 1$.
14. If $C > 0$ and $\delta > 0$, then $\|x - x_t\| \leq \delta$ (and $\|y - y_t\| \leq \delta$) for

all $t \in \mathbb{C}$ with $|t| \le C$.

A functional calculus property

15. If $0 < \alpha < \beta < 1$, then there are analytic semigroups $t \mapsto (a^{\alpha} - a^{\beta})^{t}$: $H \to A$ such that $(a^{\alpha} - a^{\beta})^{1} = a^{\alpha} - a^{\beta}$.

A subsidiary hypothesis in one of the properties, for example, of the form "if..., then..." is assumed to hold before the semigroup is chosen.

3.2 REMARKS

If the bound on the approximate identity in the hypotheses of Theorem 3.1 is d (≥ 1), then properties 1 to 4 of Theorem 3.1 still hold though Property 5 fails (see Dixon [1978] for this). This slightly stronger version of 3.1 with $d \ge 1$ is proved in 4.7. However the stronger form enables us to pass to an equivalent norm and choose a new approximate identity for which $d = 1$.

3.3 COROLLARY

A Banach algebra with a countable bounded approximate identity has a commutative bounded approximate identity bounded by 1 in an equivalent Banach algebra norm.

In the proof of Corollary 3.3 we require the following standard little renorming lemma.

3.4 LEMMA

Let S be a bounded multiplicatively closed subset (that is, an algebraic semigroup) in a Banach algebra B. Then there exists an equivalent Banach algebra norm $|\cdot|$ on B such that $|b| \le 1$ for all $b \in S$.

Proof of Lemma. Let $q(x) = \sup \{\|x\|, \|cx\| : c \in S\}$ for all $x \in B$. Straightforward calculations show that q is an algebra norm on B satisfying $\|x\| \le q(x) \le M\|x\|$ for all $x \in B$, where M is an upper bound for $\{1, \|c\| : c \in S\}$. Also $q(cx) \le q(x)$ for all $c \in S$ and all $x \in B$. Let $|b| = \sup \{q(bx) : x \in B, q(x) \le 1\}$ for all $b \in B$. Then $|\cdot|$ is the required norm on B.

Proof of Corollary 3.3. Let $t \mapsto a^t : H \to A$ be a semigroup provided by Theorem 3.1 for A. There is a real number M such that $\{\|a^t\| \exp(-Mt) : t > 0\}$ is bounded. This is because it follows from Property 3 that $\{\|a^t\| : 0 < t < 1\}$ is bounded - we could choose $M = \log \|a^1\|$. The multiplicatively closed set $\{\exp(-Mt)\ a^t : t > 0\}$ in A gives an equivalent Banach algebra norm $|\cdot|$ on A such that $|\exp(-Mt).a^t| \leq 1$ for all $t > 0$ (Lemma 3.4). Property 4 implies that $\{\exp(-Mt).a^t : t > 0\}$ is the required commutative bounded approximate identity for A.

Examples due to Dixon [1978] show that the hypothesis of countability cannot be omitted from Corollary 3.3. From now on we shall assume that our approximate identities are bounded by 1. Property 3 is essentially the best possible as there are Banach algebras B with bounded approximate identities in which we cannot have an analytic semigroup $t \mapsto a^t : H \to B$ such that $(a^t B)^- = B$ for all $t \in H$ and $\{\|a^t\| : t \in H, |t| \leq 1\}$ is bounded (see 5.14).

The convolution algebra $L^1(\mathbb{R}^+)$ plays some of the roles in the class of Banach algebras with countable bounded approximate identities bounded by 1 that the Banach algebra $\mathbb{C}$ does in the class of unital Banach algebras. This is the intuitive idea behind the next three corollaries. Corollary 3.5 corresponds for Banach algebras with a countable bounded approximate identity bounded by 1 to the following trivial observation for unital algebras. A Banach algebra B is unital if and only if there is a homomorphism θ from $\mathbb{C}$ into B such that $\|\theta\| = 1$ and $\theta(\mathbb{C}).B = B = B.\theta(\mathbb{C})$.

3.5 COROLLARY

Let A be a Banach algebra. Then A has a countable bounded approximate identity bounded by 1 if and only if there is a homomorphism θ from $L^1(\mathbb{R}^+)$ into A such that $\|\theta\| = 1$ and $\theta(L^1(\mathbb{R}^+)).A = A = A.\theta(L^1$

Proof. If θ exists, then a countable bounded approximate identity (f_n) bounded by 1 in $L^1(\mathbb{R}^+)$ is easily transferred to a countable bounded approximate identity $(\theta(f_n))$ in A.

Conversely suppose that A has a countable bounded approximate identity bounded by 1. Using Theorem 3.1 we choose a semigroup $t \mapsto a^t : H \to$ and we define $\theta : f \mapsto \int_0^\infty f(t)a^t\, dt : L^1(\mathbb{R}^+) \to A$. Direct calculations using $\|a^t\| \leq 1$ for all $t > 0$ and the convolution product of $L^1(\mathbb{R}^+)$

show that θ is a norm reducing homomorphism from $L^1(\mathbb{R}^+)$ into A. Now we regard A as a Banach $L^1(\mathbb{R}^+)$-module by defining $f.a = \theta(f)a$ and $a.f = a\theta(f)$ for all $a \in A$ and $f \in L^1(\mathbb{R}^+)$. We apply Theorem 3.1 to the Banach algebra $L^1(\mathbb{R}^+)$ and its left and right Banach $L^1(\mathbb{R}^+)$-modules A and A. Property 1 implies that the closed linear span of $L^1(\mathbb{R}^+).A$ is equal to $L^1(\mathbb{R}^+).A$, and similarly for $A.L^1(\mathbb{R}^+)$. Therefore $\theta(L^1(\mathbb{R}^+))A$ and $A\theta(L^1(\mathbb{R}^+))$ are closed linear subspaces of A. We have essentially proved a standard corollary of Cohen's factorization theorem in doing this (see Hewitt and Ross [1970], p.268). To show that these two closed linear subspaces of A are equal to A it is sufficient to prove that, if $b \in A$, then $b \in (\theta(L^1(\mathbb{R}^+))A)^-$. Let f_n be n times the characteristic function of [0,1/n]. For each positive integer n and each $b \in A$,

$$\begin{aligned} &\|b - \theta(f_n)b\| \\ &= \left\|\int_0^\infty f_n(t)\ (b - a^t b)\ dt\right\| \\ &\leq n \int_0^{1/n} \|b - a^t b\|\ dt \\ &\leq \sup\{\|b - a^t b\| : 0 < t \leq 1/n\}, \end{aligned}$$

and this tends to zero as n tends to infinity (Property 4). This proves Corollary 3.5.

If the algebra A had an identity, then the map θ would not be one-to-one because $L^1(\mathbb{R}^+)$ does not have an identity. There are homomorphisms from $L^1(\mathbb{R}^+)$ onto radical Banach algebras that are not one-to-one. For example let $J = \{f \in L^1(\mathbb{R}^+) : f = 0$ a.e. on $[0,1]\}$, and let ψ be the quotient map from $L^1(\mathbb{R}^+)$ onto $L^1(\mathbb{R}^+)/J$, which is isomorphic to $L^1_*[0,1]$ (see 2.13 for the definition of the Volterra algebra $L^1_*[0,1]$). However if A is a radical Banach algebra, and if θ is constructed as an integral with a continuous semigroup as in the proof of Corollary 3.5, then θ is one-to-one.

3.6 <u>THEOREM</u>

Let A be a radical Banach algebra, and let $t \mapsto a^t : (0,\infty) \to A$ be a continuous semigroup such that $\|a^t\| \leq 1$ and $(a^tA)^- = A = (Aa^t)^-$ for all $t > 0$. If $\theta : f \mapsto \int_0^\infty f(t)a^t dt : L^1(\mathbb{R}^+) \to A$, then θ is a monomorphism from $L^1(\mathbb{R}^+)$ into A satisfying $\|\theta\| = 1$ and $\theta(L^1(\mathbb{R}^+)).A = A = A.\theta(L^1(\mathbb{R}^+))$.

From the proof of Corollary 3.5 and the observation that $\{a^{1/n} : N \in \mathbb{N}\}$ is a bounded approximate identity in A, we see that it is sufficient to show that θ is one-to-one. This will follow easily from the following theorem of Allan [1979], which is proved in appendix A2.

3.7 THEOREM

Let ω be a radical weight on $\mathbb{R}^+$, and let I be a closed ideal in $L^1(\mathbb{R}^+,\omega)$. If there is a non-zero $f \in L^1(\mathbb{R}^+) \cap I$, then there is a $C \geq 0$ such that $I = \{g \in L^1(\mathbb{R}^+,\omega) : g = 0 \text{ a.e. on } [0,C]\}$.

Proof of Theorem 3.6. Let $\omega(t) = \|a^t\|$ for all $t > 0$. Since A is a radical Banach algebra, $\|a^n\|^{1/n} \to 0$ as $n \to \infty$. Using this and $\|a^t\| \leq 1$ for $0 < t \leq 1$, we obtain $\omega(t)^{1/t} \to 0$ as $t \to \infty$. Thus ω is a radical weight. We now extend θ to a map from $L^1(\mathbb{R}^+,\omega)$ into A by $\theta(f) = \int_0^\infty f(t)\, a^t\, dt$ for all $f \in L^1(\mathbb{R}^+,\omega)$. Suppose that $\ker\theta \cap L^1(\mathbb{R}^+) \neq \{0\}$. Since $\ker\theta$ is a closed ideal in $L^1(\mathbb{R}^+,\omega)$, there is a $C \geq 0$ such that $\ker\theta = \{g \in L^1(\mathbb{R}^+,\omega) : g = 0 \text{ a.e. on } [0,C]\}$. If g_n is the characteristic function of the closed interval $[C, C+1/n]$ for n a positive integer, then $a^C = \lim \int_0^\infty n g_n(t) a^t dt = 0$. This contradicts the hypotheses and completes the proof.

In the proof of Theorem 3.6 we used the homomorphism θ from $L^1(\mathbb{R}^+,\omega)$ but only proved it is one-to-one on $L^1(\mathbb{R}^+)$.

3.8 PROBLEM

Let A be a commutative radical Banach algebra, let $t \mapsto a^t : (0,\infty) \to A$ be a continuous (bounded) semigroup such that $(a^tA)^- = A$ for all $t > 0$, and let $\omega(t) = \|a^t\|$ for all $t > 0$. If $\theta : f \mapsto \int_0^\infty f(t) a^t dt : L^1(\mathbb{R}^+,\omega) \to A$, is θ one-to-one?

Property 8 ensures that by suitably choosing the semigroup $t \mapsto a^t : H \to A$ the conclusion of Theorem 3.6 holds without the hypothesis that the algebra is radical.

Property 6 implies that the function $t \mapsto a^{t+1} : H^- \to A$ is of exponential type in the closed half plane H^- because $\|a^{1+t}\| \le 2^{1/2}C_{1/2}$ $e^{\pi|\mathrm{Im}\, t|/2} \le 2^{1/2}C_{1/2}\, e^{\pi|t|/2}$ for all $t \in H^-$ (we are taking $\alpha = 1/2$). Note that the order of growth of $\|a^{x+iy}\|$ for fixed $x > \alpha > 0$ as $|y|$ tends to infinity is that of the fractional integral semigroup in $L^1(\mathbb{R}^+)$. In Chapter 5 we shall observe that a substantial improvement on the size of $\|a^t\|$ is not possible in general.

Property 7 is essentially the best possible. There are commutative Banach algebras with a countable bounded approximate identity such that the spectrum of each element in the algebra is the closure of the interior of the spectrum; for example, the maximal ideal of functions vanishing at 1 in the disc algebra has this property.

We shall apply property 9 to certain group algebras and algebras of compact operators on suitable separable nuclear C^*-algebras.

3.9 <u>COROLLARY</u>

Let G be a locally compact group. Then G is metrizable if and only if there is an analytic semigroup $t \mapsto a^t : H \to L^1(G)$ such that $\|a^t\| = 1$, such that $a^t \ge 0$ as a function and as an element of the $*$-algebra $L^1(G)$ for all $t > 0$, and such that $\|a^t * f - f\|_1 +$ $\|f * a^t - f\|_1 \to 0$ as $t \to 0$ non-tangentially in H for each $f \in L^1(G)$.

<u>Proof</u>. Suppose that G is a metrizable locally compact group. The metrizability of the topology of G implies that there is a countable open base for the topology at the identity of G. This countable base of the topology at the identity gives a countable bounded approximate identity Λ_1 in $L^1(G)$. If U_n is one of the countable open base sets, which may be assumed to have compact closure, then $f_n = \mu(U_n)^{-1}\chi_{U_n}$ is a typical element of Λ_1, where χ_E denotes the characteristic function of E and μ is left Haar measure (see the proof of Theorem 20.27 of Hewitt and Ross [1963] p.303 for further details). The elements of Λ_1 are bounded by 1 in $\|\cdot\|_1$ - norm and are positive as functions. The involution on $L^1(G)$ is isometric and preserves the positivity of functions, and convolution preserves the positivity of functions. Thus the set $\Lambda_2 = \{g^* * g : g \in \Lambda_1\}$ is a countable bounded approximate identity in $L^1(G)$ bounded by 1 consisting of positive functions. Let Λ denote the set of $g \in L^1(G)$ such that $g \ge 0$ almost everywhere on G, g is self-adjoint in the $*$-algebra $L^1(G)$, and $\|g\|_1 \le 1$. Since $\Lambda \supseteq \Lambda_2$, the set Λ is a bounded approximate identity

in $L^1(G)$. Also Λ is closed convex and $g^n \in \Lambda$ if $g \in \Lambda$ and $n \in \mathbb{N}$, where g^n is the n-th convolution power of g.

We apply Theorem 3.1 to the Banach algebra $L^1(G)$ and the bounded approximate identity Λ in it. This gives an analytic semigroup $t \mapsto b^t : H \to L^1(G)$ such that $b^t \in \Lambda$ for all $t > 0$, and $\|b^t * f - f\|_1 + \|f * b^t - f\|_1 \to 0$ as $t \to 0$ in any sector $\{z \in \mathbb{C} : z \neq 0, |\text{Arg } z| < \psi\}$ with $0 < \psi < \pi/2$. Let $a^t = b^t (\int_G b^t \, d\mu)^{-1}$ for all $t \in H$. Since the function $f \mapsto \int_G f d\mu : L^1(G) \to \mathbb{C}$ is a character, $t \mapsto a^t : H \to L^1(G)$ is an analytic semigroup. The order properties of a^t for $t > 0$ follow from the definition of Λ and $a^t = (a^{t/2})^* * a^{t/2}$.

Conversely suppose that the semigroup exists. Let U be a compact neighbourhood of the identity in G, and let $\psi : x \mapsto \delta_x * a^1 : U \to L^1(G)$, where δ_x is the point mass at $x \in G$. Note that $\delta_x * a^1$ is a^1 shifted by x in $L^1(G)$. Then ψ is a continuous function from U with the relative topology of G into $L^1(G)$ (by Hewitt and Ross [1963] p.285 since $\delta_x * a^1 = a^1_{x^{-1}}$ in the notation of Hewitt and Ross [1963] p.285). Now U is compact and $L^1(G)$ is Hausdorff so ψ will be a homeomorphism from U onto $\psi(U)$ if ψ is one-to-one. If $\delta_x * a^1 = \delta_y * a^1$, then $(\delta_x - \delta_y) * L^1(G) \subseteq ((\delta_x - \delta_y) * a^1 * L^1(G))^- = \{0\}$ so that $\delta_x - \delta_y = 0$ and $x = y$. Thus ψ is a homeomorphism and the topology of G is metrizable.

We shall use Theorem 3.1 to show the existence of semigroups of completely positive compact operators on suitable nuclear C^*-algebras. It will be clear from the proof that analogous results hold for suitable Banach spaces satisfying the metric approximation property (for example, if the Banach space and its dual are separable and satisfy the metric approximation property). However in the Banach space case the order properties are lost. We start by recalling the definitions of a completely positive operator and of a nuclear C^*-algebra.

If X is a Banach space, let $CL(X)$ denote the Banach algebra of compact linear operators on X, and let $FL(X)$ denote the algebra of continuous finite rank linear operators on X. An operator T on a C^*-algebra B is said to be positive if $Tb \geq 0$ for all $b \geq 0$. Let $M_n(B)$ denote the C^*-algebra of $n \times n$ matrices with entries from the C^*-algebra B, and let I_n denote the identity operator from the C^*-algebra $M_n(\mathbb{C})$ onto $M_n(\mathbb{C})$. We shall think of $M_n(B)$ as $M_n(\mathbb{C}) \otimes B$. An operator T on a C^*-algebra B, is said to be completely positive if $T \otimes I_n$ is a positive operator on $M_n(B)$ for all $n \in \mathbb{N}$. One of the equivalent formulations that a C^*-algebra is nuclear is that $CL(B)$ has a left bounded approximate identity bounded by 1 consisting of completely positive continuous finite rank operators (see Lance [1973] and Choi and Effros [1978]). On a commutative C^*-algebra each positive operator is completely positive (Stinespring [1955]).

3.10 <u>COROLLARY</u>

Let B be a separable nuclear C^*-algebra, and suppose that $CL(B)$ has a bounded approximate identity of completely positive (continuous) finite rank operators bounded by 1. Then for each separable subspace Y of $CL(B)$ there is an analytic semigroup $t \mapsto a^t : H \to CL(B)$ such that $\|a^t\| \leq 1$ and a^t is completely positive for each $t > 0$, and that $\|a^t.T - T\| + \|R.a^t - R\| \to 0$ as $t \to 0$ non-tangentially in H for all $T \in CL(B)$ and all $R \in Y$.

<u>Proof</u>. We require a compact linear operator T_0 on B such that $(T_0.CL(B))^- = CL(B)$. There are several ways to obtain this from the hypotheses. Here is one way. Let $\{x_n \in B : n \in \mathbb{N}\}$ be a countable dense subset in the unit ball of B. By regarding B as a left Banach $CL(B)$-module, the sequence (x_n) in B may be factored by the Varopoulos-Johnson version of Cohen's factorization theorem (see Hewitt and Ross [1970], p.268) in the form $T_0y_n = x_n$ for $y_n \in B$ and $T_0 \in CL(B)$. Hence $(T_0.B)^- = B$. If $f \in B^*$ and $x \in B$, then $T_0.(f \otimes x) = f \otimes T_0x$. From this it follows that $(T_0.CL(B))^- \supseteq FL(B)$. Because $FL(B)^- = CL(B)$, it follows that $(T_0.CL(B))^- = CL(B)$.

By induction on n we choose a countable bounded approximate identity E_n of completely positive finite rank operators bounded by 1 acting as an approximate identity for T_0, Y, and themselves (that is, $\|E_nE_m - E_m\| + \|E_mE_n - E_m\| \to 0$ as $n \to \infty$ for each m). In making this choice we use the separability of Y to ensure we can choose a sequence to act as a right approximate identity for Y. Let A be the closed subalgebra of $CL(B)$ generated by $\{E_n : n \in \mathbb{N}\}$, T_0, and Y. Then the sequence (E_n) is a countable bounded approximate identity for A. We apply Theorem 3.1 to A with Λ the set of completely positive operators in A with norm less than or equal to 1, and with $X = CL(B)$ and $Y = Y$. Since the product of two completely positive operators is completely positive, Λ satisfies the hypotheses of Property 9. This completes the proof.

3.11 PROBLEM

Which separable nuclear C^*-algebras B have the property that $CL(B)$ has a bounded approximate identity of completely positive finite ra[nk] operators bounded by 1?

The following theorem due to Johnson [1972], Lemma 6.2, shows that certain commutative C^*-algebras satisfy the hypotheses of Corollary 3.10, because a positive operator on a commutative C^*-algebra is completel[y] positive.

3.12 THEOREM

Let Ω be a compact Hausdorff space. If there is a positive regular Borel measure λ on Ω with the support of λ equal to Ω, the[n] $CL(C(\Omega))$ has a bounded approximate identity of finite rank positive operators bounded by 1.

Proof. Let $\mathcal{G}$ denote the set of partitions of unity $G = \{g_1,\ldots,g_n\}$ in $C(\Omega)$ such that $0 \le g_j \le 1$, each g_j is non-zero, and $\sum_1^n g_j = 1$. For each $G \in \mathcal{G}$ and each regular Borel measure μ on Ω with $\mu \ge \lambda$, we let $E_{G,\mu} \in CL(C(\Omega))$ be defined by

$$E_{G,\mu} f = \sum_1^n \langle fg_j,\mu\rangle \langle g_j,\mu\rangle^{-1} g_j$$

for all $f \in C(\Omega)$. Note that $\langle\cdot,\cdot\rangle$ is the pairing between $C(\Omega)$ and its dual $M(\Omega)$, the space of regular Borel measures, and that since μ is positive with support Ω, $\langle g_j,\mu\rangle \ne 0$ for all j. Direct calculations using the positivity of μ show that $E_{G,\mu}$ is a positive finite rank operator, and that $\|E_{G,\mu}\| \le 1$. We shall now show that $\{E_{G,\mu} : G \in \mathcal{G},\ \mu \in M(\Omega),\ \mu \ge \lambda\}$ is the required bounded approximate identity for $CL(C(\Omega))$.

Since each $E_{G,\mu}$ satisfies $\|E_{G,\mu}\| \le 1$ and since $FL(C(\Omega))$ i[s] dense in $CL(C(\Omega))$, it is sufficient to show that $\{E_{G,\mu} : G \in \mathcal{G},\ \mu \in M(\Omega),\ \mu \ge \lambda\}$ is a bounded approximate identity for the continuous finite rank operators. Each continuous finite rank operator T on $C(\Omega)$ may be written $T = \sum_1^n f_j \otimes \mu_j$, where $f_j \in C(\Omega)$ and $\mu_j \in M(\Omega)$

Calculating the two products $E_{G,\mu}T$ and $TE_{G,\mu}$, we see that it is sufficient to prove that for each $\varepsilon > 0$, each $f_1,\ldots,f_n \in C(\Omega)$, and each $\mu_1,\ldots,\mu_n \in M(\Omega)$ there is a $G \in \mathcal{G}$ and a $\mu \geq \lambda$ such that $\|E_{G,\mu} f_j - f_j\| < \varepsilon$ and $\|E^*_{G,\mu} \mu_j - \mu_j\| < \varepsilon$ for $1 \leq j \leq n$. Here $E^* : M(\Omega) \to M(\Omega)$ is the adjoint of E.

Let $\mu = |\mu_1| + \ldots + |\mu_n| + \lambda$. Then $\mu \geq \lambda$ and $\mu \geq |\mu_j|$ for each j so by the Radon-Nikodym Theorem there are $\phi_1,\ldots,\phi_n \in L^1(\mu)$ such that $\mu_j = \phi_j\mu$ for $1 \leq j \leq n$. Since $C(\Omega)$ is dense in $L^1(\mu)$ in $\|\cdot\|_1$, there are $h_1,\ldots,h_n \in C(\Omega)$ such that $\|\phi_j - h_j\|_1 < \varepsilon/4$ for $1 \leq j \leq n$. Now $L^1(\mu)$ is a closed invariant subspace in $M(\Omega)$ for $E^*_{G,\mu}$ for all $G \in \mathcal{G}$. Further

$$\begin{aligned}
&\|E^*_{G,\mu} \mu_j - \mu_j\| \\
&= \|E^*_{G,\mu} \phi_j - \phi_j\|_1 \\
&\leq \varepsilon/2 + \|E^*_{G,\mu} h_j - h_j\|_1 \\
&\leq \varepsilon/2 + \| E_{G,\mu} h_j - h_j\|_\infty . \mu(\Omega)
\end{aligned}$$

for all j. We must thus choose G so that $\|E_{G,\mu} h - h\|_\infty < \nu = \min\{\varepsilon/2, \varepsilon/2 . \mu(\Omega)^{-1}\}$ for all $h \in \{f_1,\ldots,f_n, h_1,\ldots,h_n\} = K$. For each j, $h \mapsto \langle hg_j,\mu\rangle\langle g_j,\mu\rangle^{-1}$ defines a probability integral on $C(\Omega)$ and thus $\langle hg_j,\mu\rangle\langle g_j,\mu\rangle^{-1}$ is in the closed convex hull of $\{h(w) : w \in \operatorname{supp} g_j\}$ so that

$$\begin{aligned}
&|\langle hg_j,\mu\rangle\langle g_j,\mu\rangle^{-1} - h(x)| \\
&\leq \sup \{|h(y) - h(x)| : y \in \operatorname{supp} g_j\}
\end{aligned}$$

for all $x \in \Omega$. For each $h \in C(\Omega)$ and $x \in \Omega$, we have

$$\begin{aligned}
&|E_{g,\mu}h(x) - h(x)| \\
&\leq \sum_1^m |\langle hg_j,\mu\rangle\langle g_j,\mu\rangle^{-1} - h(x)| \; g_j(x)
\end{aligned}$$

since $G = \{g_1,\ldots,g_m\}$ is a partition of unity in $C(\Omega)$. Hence

$$|E_{G,\mu}\, h(x) - h(x)|$$

$$\leq \sum_{1}^{m} \sup\{|h(w) - h(y)| : w,y \in \text{supp}\, g_j\}\ \ g_j(x)$$

$$\leq \max\,\{\sup\{|h(w) - h(y)| : w,y \in \text{supp}\, g_j\} : 1 \leq j \leq m\}.$$

Since K is a finite set, we can find a partition of unity G in $C(\Omega)$ such that $\sup\{|h(w) - h(y)| : w,y \in \text{supp}\, g_j\} < \nu$ for $1 \leq j \leq m$ and $h \in K$. This completes the proof.

Property 13 shows that $t \mapsto x_t : \mathbb{C} \to X$ may be chosen to be an entire function bounded above in norm by a constant times $\exp(|t|^{1+\delta})$ for $\delta > 0$. In general the function $t \mapsto x_t$ is not of exponential type as the following argument shows. Let A be a commutative radical Banach algebra, and suppose that a non-zero element $x \in X$ can be factored $x = a^t.x_t$ with $t \mapsto a^t : H \to A$ an analytic semigroup and $t \mapsto x_t : H \to X$ an analytic functi of exponential type. Then there are positive constants K_1 and K_2 such that $\|x_t\| \leq K_1 \exp(K_2|t|)$ for all $t \in H$. We now have $\|x\| \leq \|a^t\|\, K_1 \exp(K_2|t|)$ for all $t \in H$ so that $\lim_{t\to\infty} \|a^t\|^{1/t} \geq \exp(-K_2) > 0$. This contradicts A being a radical Banach algebra.

Here is another result obtained by a variation of this argument.

3.13 <u>COROLLARY</u>

Let A be a Banach algebra with a countable bounded approximate identity (bounded by 1). If $t \mapsto \gamma_t : [0,\infty) \to (0,1)$ is a continuous function such that $\gamma_t \to 0$ as $t \to \infty$, then there is an analytic semigroup $t \mapsto a^t : H \to A$ such that $\|a^t\|^{1/|t|} \geq \gamma_{|t|}$ for all $t \in H$ with $|t| \geq 1$.

<u>Proof</u>. Let $\alpha_t = \gamma_t^{-1}$ for all $t \in [0,\infty)$. Then $t \mapsto \alpha_t$ is a continuous function from $[0,\infty)$ into $(1,\infty)$ with $\alpha_t \to \infty$ as $t \to \infty$, and hence $\alpha_t \geq 1 + \delta$ for some $\delta > 0$ and all $t \geq 0$. Let x be an element of A with $\|x\| = 1$, and factorize x as in Theorem 3.1. Then $\|x_t\| \leq (\alpha_{|t|})^{|t|}\|x$ for all $t \in \mathbb{C}$ with $|t| \geq 1$ so that

$\|a^t\| \geq \|x_t\|^{-1} \geq (\gamma_{|t|})^{|t|}$ for all $t \in H$ with $|t| \geq 1$.

This corollary says that in a radical Banach algebra with a bounded approximate identity there are analytic semigroups with $\|a^t\|^{1/|t|}$

tending to zero arbitrarily slowly as $|t|$ tends to infinity with $t \in H$. In 5.3 we shall see that $\|a^t\|^{1/t}$ cannot tend to zero arbitrarily fast as t tends to infinity.

We know from the analyticity of $t \mapsto x_t : \mathbb{C} \to X$ that x_t tends to x as t tends to zero. Property 14 says that for a preassigned δ and bounded region of $\mathbb{C}$ we can ensure that $\|x - x_t\| < \delta$.

The only motivation that I have for extracting property 15 is that it gives the Banach algebra generalization of a type of bounded approximate identity that has played a crucial role in two deep results in C^*-algebras (see Arveson [1977] and Elliott [1977]). We shall briefly define the type of approximate identity obtained from property 15 but we shall not consider its use in C^*-algebras. Let (λ_n) be a strictly decreasing sequence of positive real numbers converging to zero with $\lambda_1 < 1$, and let $E_o = a^{\lambda_1/2}$ and $E_n = (a^{\lambda_{n+1}} - a^{\lambda_n})^{1/2}$ for all $n \in \mathbb{N}$. Then the sequence $\sum_1^n E_j^2$ is a bounded approximate identity for A. It is this form of the approximate identity together with the special order properties of C^*-algebras inherited by the sequence (E_n) that are used in Arveson [1977] and Elliott [1977].

Let B be a Banach algebra containing the Banach algebra A as a closed ideal. Then A is said to have a bounded approximate identity quasicentral for B if for each finite subset F of A, each finite subset K of B, and each $\varepsilon > 0$, there is an $e \in A$ with $\|e\| \le 1$, and $\|ea - a\| + \|ae - a\| < \varepsilon$ for all $a \in F$, and $\|be - eb\| < \varepsilon$ for all $b \in K$. This definition can easily be translated into one about nets. In appendix A3 we show that an Arens regular Banach algebra A with a bounded approximate identity has a quasicentral bounded approximate identity for all enveloping algebras of A. However the following problem seems to be open.

3.14 <u>PROBLEM</u>

Let A be a Banach algebra with a bounded approximate identity bounded by 1, and let $\mathrm{Mul}(A)$ be the multiplier algebra of A. When does A have a bounded approximate identity that is quasicentral for $\mathrm{Mul}(A)$?

If the bounded approximate identity in the Banach algebra A has nice properties with respect to a suitable set of derivations, multipliers or automorphisms, then these properties may be inherited by a^t as

$t \to 0$ by suitably choosing a^t. This is the intuitive idea behind Theorem 3.15 which continues the properties of Theorem 3.1.

3.15 THEOREM

Let A be a Banach algebra with a countable bounded approximat identity bounded by 1 and without identity. Then an analytic semigroup $t \mapsto a^t : H \to A$ may be chosen so that properties 1 to 5 and 7 to 14 of Theorem 3.1 hold, and that one of the following properties hold.

16. Let Z be a separable Banach space of continuous derivations on A. If there is a bounded approximate identity $(g_n) \subseteq \Lambda$ satisfying

    $\|D(g_n)\| \to 0$ as $n \to \infty$ for all $D \in Z$, then $\|D(a^t)\| \to 0$ as $t \to 0$, $t > 0$, for all $D \in Z$.

17. If B is a Banach algebra containing A as a closed ideal, if B/A is separable, and if A has a bounded approximate identity in Λ quasicentral for B, then $\|ba^t - a^t b\| \to 0$ as $t \to 0$, $t > 0$, for all $b \in B$.

18. Let G be a group of continuous automorphisms of A, and suppose tha G contains a countable dense subset (in the uniform norm topology). If there is a bounded approximate identity $(g_n) \subseteq \Lambda$ in A satisfyin $\|\beta(g_n) - g_n\| \to 0$ as $n \to \infty$ for all $\beta \in G$, then $\|\beta(a^t) - a^t\| \to 0$ as $t \to 0$, $t > 0$, for all $\beta \in G$.

3.16 NOTES AND REMARKS

Most of the properties in Theorems 3.1 and 3.15 are in Sinclair [1978], [1979a] but in several cases the results are only implicitly there. For example, property 3 is proved in Sinclair [1978] for the interval (0,1] in place of the sector $U(\psi)$, and I first saw the sector result in Esterle' U.C.L.A. lecture course. We shall prove the properties using the exponentia methods of Sinclair [1978] except that we shall deduce 6 and 15 from the functional calculus results of Sinclair [1979a]. Property 15 is essentially a functional calculus property but 6 is not. I do not know how to prove 6 using exponential methods, and the functional calculus factorization is not proved in these notes. The proofs of Theorems 3.1 and 3.15 are discussed in detail in Chapter 4, and are variations and extensions of Cohen's factoriza- tion theorem. There are good accounts of various forms of Cohen's factoriza- tion theorem in Hewitt and Ross [1970], Bonsall and Duncan [1973], and Doran and Wichman [1979]. The latter notes contain a detailed account of bounded approximate identities and factorization of elements in Banach modules including the results of Sinclair [1978], [1979a] with little modification. Neither of the books nor lecture notes touch the η_1-factorizations of Esterle [1978], [1980b], and Sinclair [1979b].

Corollary 3.3 was first proved for a commutative Banach algebra by Dixon [1973]. The form here is in Dixon [1978] and Sinclair [1979a]. Aarnes and Kadison [1969] showed that a separable C^*-algebra has a commutative bounded approximate identity, and Hulanicki and Pytlik [1972] (see also Pytlik [1975]) proved that $L^1(G)$ has a commutative bounded approximate identity.

Property 8 seems to be new.

Properties 7 and 9 were first proved using the functional calculus methods (Sinclair [1979]) but here are proved using exponential methods. Corollary 3.9 is due to Hunt [1956] (see Stein [1970] Chapter III) for G a connected Lie group. There have been some investigations of semi-groups of completely positive operators on C^*-algebras - see Evans and Lewis [1977], for example. Theorem 3.12 is proved for Ω the circle $\mathbb{T}$ in Johnson [1970], but his proof gives what we state here. See also Herbert and Lacey [1968].

Corollary 3.13 is the semigroup version of a result in Allan and Sinclair [1976] that in a radical Banach algebra with a one sided bounded approximate identity there are elements a with $\|a^n\|^{1/n}$ tending to zero arbitrarily slowly. Rates of growth of $\|a^n\|^{1/n}$ and semigroups in radical Banach algebras are discussed in Bade and Dales [1980], Esterle [1980], and Grønbaek [1980].

Quasicentral bounded approximate identities in C^*-algebras are used in Arveson [1976], Akermann and Pedersen [1978], and Elliott [1977]. See also A3.

4 PROOF OF THE EXISTENCE OF ANALYTIC SEMIGROUPS

In this chapter we shall prove Theorems 3.1 and 3.15, and the various lemmas required in the proofs. In 4.1 we sketch the ideas behind the proofs, and after proving all the lemmas we prove 3.1 in 4.7 and 3.15 in 4.8. Throughout this chapter A will denote a Banach algebra with a countable bounded approximate identity bounded by $d(\geq 1)$, X will denote a left Banach A-module satisfying $\|a.x\| \leq \|a\|.\|x\|$ for all $a \in A$ and $x \in X$, and $[A.X]^-$ will denote the closed linear span of the set $\{a.x : a \in A, x \in X\}$. Taking $d = 1$ simplifies the calculations slightly. We assume that A does not have an identity.

4.1 SKETCH OF THE PROOF

The proof is a variation of Cohen's factorization theorem (Cohen [1959]) with the analytic semigroup obtained as a limit of exponential semigroups in the unital Banach algebra $A^\#$. The variation is influenced by the proof of the Hille-Yoshida Theorem. If the algebra A ha an identity, then the factorization results would be trivial as we could take $a^t = 1$ for all $t \in H$. Though our algebra does not have an identity we shall use the case when there is an identity and an approximation to prove 3.1. We work in the algebra $A^\# = A \oplus \mathbb{C}1$ obtained by adjoining an identity to A, and we regard X and Y as left and right Banach $A^\#$-modules by defining $1.w = w$ and $u.1 = u$ for all $w \in X$ and $u \in Y$. In the proof we choose inductively a sequence (e_n) from the bounded approximate identity Λ of A such that

$$b_n^{\ t} = \exp(t \sum_1^n (e_j - 1)) \in A^\#$$

converges to an element $a^t \in A$ for all $t \in H$. With suitable restriction on the inductive choice of the sequence (e_n), the sequence $(b_n^{\ -t}.x)$ converges to an element x_t in X for all $t \in \mathbb{C}$. The analyticity of the functions $t \mapsto a^t : H \to A$ and $t \mapsto x_t : \mathbb{C} \to X$ follow from the calculations that show that a^t and x_t exist. In a Banach algebra the sequence $(b_n^{\,t})$ usually does not converge and $(b_n^{\,-t}.x)$ should diverge rather badly as n tends to infinity, but this does not occur with the correct choice of the sequence e_n because $(e_n - 1)$ acts on x and $e_1,\ldots,e_{n-1}$ like an approximate zero divisor. That $(b_n^{\,t})$ converges to an element in A rather than in $A\#$ is because $b_n^{\,t} \in \exp(-nt) \oplus A$ and $\exp(-nt) \to 0$ as $n \to \infty$ for all $t \in H$. The crucial semigroup property of $t \mapsto a^t$ is obtained from the corresponding semigroup property of the exponential groups $t \mapsto b_n^{\,t}$ in the unital Banach algebra A. To obtain further properties of a^t or x_t one simply imposes further restrictions on the choice of the sequence (e_n), and has calculations linking e_n with a^t and x_t through $b_n^{\,t}$ and $b_n^{\,-t}.x$. Convexity and convex combinations play a vital role in the proof, and the semigroup a^t for $t > 0$, is essentially just a weighted average of a suitable approximate identity. The averaging smooths the given approximate identity into a nicer one. This idea is clearly illustrated in the proofs of Lemma 4.5 and property 9 of Theorem 3.1. The $b_n^{\,t}$ corresponds to $\exp t(\lambda^2(\lambda - R)^{-1} - \lambda)$ in the proof of the Hille-Yoshida Theorem (6.7), and heuristically $n^{-1}(e_1 + \ldots + e_n - n)$ is approaching the infinitesimal generator of the semigroup $t \mapsto a^t$.

The first lemma is the standard opening to the stronger forms of Cohen's factorization theorem.

4.2 <u>LEMMA</u>

If w is in the closed linear span of $A.X$ and if $\varepsilon > 0$, then there are $a_1,\ldots,a_m \in A$ and $\delta > 0$ such that $\|ew - w\| < \varepsilon$ for each $e \in A$ with $\|e\| \le d$ and $\|ea_j - a_j\| < \delta$ for all j.

<u>Proof</u>. There are $a_1,\ldots,a_m \in A$ and $w_1,\ldots,w_m \in X$ so that

$$\|w - \sum_1^m a_j w_j\| < \varepsilon/3d.$$

Thus
$$\|ew - w\|$$
$$\le \sum_1^m \|ea_j - a_j\| \, \|w_j\| + (\|e\| + 1)\|\, w - \sum_1^m a_j\, w_j\|$$

for all $e \in A$. We may now choose δ to give the result.

When the algebra is non-commutative Lemma 4.3 is required in the proof of 4.4 : however 4.3 is not required if the algebra is commutati Lemma 4.3 is used to get around the failure of the formula $\exp(a+b) = \exp a.\exp b$ in A if A is non-commutative. This formula does hold if a and b commute.

4.3 LEMMA

If $f \in A$ and if $\eta = \|f\| + d + 1$, then

(a) $\|(f + e - 1)^k - f^k - (e - 1)^k\|$
$\leq \eta^k \{\|(e - 1)f\| + \|f(e - 1)\|\}$, and

(b) $\|(f + e - 1)^k.w - f^k.w\|$
$\leq \eta^k\{\|(e - 1)f\|.\|w\| + \|f(e - 1)\|.\|w\| + \|(e - 1).w\|\}$

for all $e \in A$ with $\|e\| \leq d$, all $k \in \mathbb{N}$, and all $w \in X$.

Proof. We multiply out the power $(f + (e - 1))^k$, remembering that f and $(e - 1)$ do not commute, and after taking f^k and $(e - 1)^k$ over to the same side as $(f + (e - 1))^k$ use the norm inequalities to obtain

$$\|(f + (e - 1))^k - f^k - (e - 1)^k\|$$
$$\leq \sum_{1}^{k-1} \|f\|^{k-1-j} \|e - 1\|^{j-1} \left(\binom{k}{j} - 1\right)\{\|(e - 1)f\| + \|f(e - 1)\|\}$$

We now have

$$\sum_{1}^{k-1} \|f\|^{k-1-j} \|e-1\|^{j-1} \left(\binom{k}{j} - 1\right)$$
$$\leq \sum_{1}^{k-1} \|f\|^{k-1-j} (d+1)^{j-1} \binom{k}{j}$$
$$\leq (\|f\| + d + 1)^k.$$

This proves (a).

From (a),

$$\|(f + e - 1)^k.w - f^k.w\|$$
$$\leq \|(e - 1)^k.w\| + \eta^k\{\|(e - 1)f\|.\|w\| + \|f(e - 1)\|.\|w\|\}$$

$$\leq \eta^k \{\|(e - 1).w\| + \|(e - 1)f\|.\|w\| + \|f(e - 1)\|.\|w\|\}.$$

The following is the main technical lemma used in the proof of Theorem 3.1.

4.4 LEMMA

Let $f + \mu 1 \in A^{\#}$ with $f \in A$, and let $\eta = \|f\| + d + 1$.

(a) Then

$$\|\exp t(f + e - 1) - \exp tf - \exp t(e - 1) + 1\|$$
$$\leq (\exp(\eta|t|) - 1)(\|(e - 1)f\| + \|f(e - 1)\|)$$

for all $e \in A$ with $\|e\| \leq d$ and all $t \in \mathbb{C}$.

If $w \in X$ and $M > 0$, then there are constants ζ and ξ depending on $\|f\|$, $|\mu|$, M, and d such that

(b) $$\|\exp t(f + \mu 1 + e - 1) - \exp t(f + \mu 1)\|$$
$$\leq \exp \mathrm{Re}(t\mu) \,.\, \{\exp|t|(d+1) - 1 + \zeta(\|(e - 1)f\| + \|f(e - 1)\|)\},$$

and

(c) $$\|\exp t(f + \mu 1 + e - 1).w - \exp t(f + \mu 1).w\|$$
$$\leq \xi\{\|(e - 1)f\|.\|w\| + \|f(e - 1)\|.\|w\| + \|(e - 1).w\|\}$$

for all $t \in \mathbb{C}$ with $|t| \leq M$ and all $e \in A$ with $\|e\| \leq d$.

Proof. Expanding the three exponentials in power series in the algebra A, we obtain

$$\|\exp t(f + e - 1) - \exp tf - \exp t(e - 1) + 1\|$$
$$\leq \sum_{k=1}^{\infty} \frac{|t|^k}{k!} \|(f + e - 1)^k - (e - 1)^k - f^k\|$$
$$\leq \sum_{k=1}^{\infty} \frac{|t|^k}{k!} \eta^k\{\|(e - 1)f\| + \|f(e - 1)\|\}$$

by Lemma 4.3. This proves (a)

Using (a) we have

$$\|\exp t(f + \mu 1 + e - 1) - \exp t(f + \mu 1)\|$$
$$= \exp \mathrm{Re}(t\mu).\|\exp t(f + e - 1) - \exp tf\|$$
$$\leq \exp \mathrm{Re}(t\mu)\{\|\exp t(e - 1) - 1\| + (\exp(|t|\mu) - 1).(\|(e - 1)f\| + \|f(e - 1)\|)\}$$

$$\leq \exp \mathrm{Re}(t\mu)\ \{\exp|t|(d + 1) - 1 + \zeta(||(e - 1)f|| + ||f(e - 1)||)$$

with $\zeta = \exp(M\eta) - 1$. We have used the inequality

$$\begin{aligned} &||\exp t(e - 1) - 1|| \\ &\leq \exp(|t|||e - 1||) - 1 \\ &\leq \exp(|t|(d + 1)) - 1. \end{aligned}$$

This proves (b).

There is a proof of (c) similar to that of (b) using (a). However we proceed by expanding the exponentials in series as in (a) to obtai

$$\begin{aligned} &||\exp t(f + \mu 1 + e - 1).w - \exp t(f + \mu 1).w|| \\ &\leq \exp \mathrm{Re}(t\mu).\ \sum_{k=1}^{\infty} \frac{|t|^k}{k!} ||(f + e - 1)^k.w - f^k.w|| \\ &\leq (\exp M|\mu|).\ \sum_{k=1}^{\infty} \frac{M^k\eta^k}{k!} \{||(e - 1)f||.||w|| + ||f(e - 1)||.||w|| + \\ &||(e - 1).w||\} \end{aligned}$$

for all $t \in \mathbb{C}$ with $t \leq M$ and all $e \in A$ with $||e|| \leq d$. We choose $\zeta = \exp M(\eta + |\mu|)$. This proves Lemma 4.4.

We now sketch a proof of the commutative case of Lemma 4.4(a) avoiding the use of 4.3. If f and e commute, then

$$\begin{aligned} &||\exp t(f + e - 1) - \exp tf - \exp t(e - 1) + 1|| \\ &= ||(\exp t(e - 1) - 1)(\exp tf - 1)|| \\ &\leq \sum_{k=1}^{\infty} \frac{|t|^k}{k!} ||e - 1||^{k-1} ||(e - 1)(\exp tf - 1)|| \\ &\leq (\exp(|t|(d + 1)) - 1).(\exp|t|||f|| - 1).||(e - 1)f|| \end{aligned}$$

for all $t \in \mathbb{C}$. This is an estimate like 4.4(a), and may be used in place of 4.4(a) to deduce (b) and (c).

The following lemma is used in the proof of Property 7 of Theorem 3.1, and provides control of the spectrum of a bounded approximate identity.

4.5 LEMMA

Let A be a commutative Banach algebra with a bounded approximate identity Λ. If Λ is convex, then given $a_1, \ldots, a_m \in A$ and $\alpha > 0$, there is $f \in \Lambda$ such that $\|fa_j - a_j\| \le \alpha$ for $1 \le j \le m$ and $|\mathrm{Im}\, \sigma(f)| \le \alpha$, where $\sigma(f)$ is the spectrum of f.

Proof. Let $\Lambda_o = \{g \in \Lambda : \|ga_j - a_j\| \le \alpha$ for $1 \le j \le m\}$. Then Λ_o is a convex bounded approximate identity for A bounded by d (say). We choose a positive integer n with

$$n > 2d\alpha^{-1}, \tag{1}$$

and by induction $e_1, \ldots, e_n \in \Lambda_o$ such that

$$\|e_k e_j - e_j\| \le \alpha^2/4 \text{ for } 1 \le j < k \le n. \tag{2}$$

Let $f = n^{-1}(e_1 + \ldots + e_n)$. Then $f \in \Lambda$ and $\|fa_j - a_j\| \le \alpha$ for $1 \le j \le n$ since Λ_o is convex. We check that f has the required spectral property by showing that if $\phi \in \Phi$, the carrier space of A, then $|\mathrm{Im}\phi(f)| \le \alpha$. If $\phi \in \Phi$ and $1 \le k < n$, then either

$$|\phi(e_k) - 1| \le \alpha/2 \quad \text{or}$$
$$|\phi(e_j)| \le \alpha/2 \quad \text{for } 1 \le j < k.$$

If both of these inequalities fail, then $\|e_k e_j - e_j\| \ge |\phi(e_j)|.|\phi(e_k) - 1| > \alpha^2/4$, which contradicts inequality (2). Now $|\phi(f)| \le \alpha/2 < \alpha$ provided $|\phi(e_j)| \le \alpha/2$ for $1 \le j \le n$. So suppose that there exists j such that

$$|\phi(e_k)| \le \alpha/2 \text{ for } 1 \le k < j \text{ and that } |\phi(e_j)| > \alpha/2. \tag{3}$$

Then

$$|\phi(e_k) - 1| \le \alpha/2 \quad \text{for } j < k \le n, \tag{4}$$

and so $|\mathrm{Im}\, \phi(e_k)| \le \alpha/2$ for $k \ne j$ and $1 \le k \le n$. Let $w = (n - 1)^{-1}(e_1 + \ldots + e_{j-1} + e_{j+1} + \ldots + e_n)$. So w is a convex combination of the e_k $(k \ne j)$ and $\|w\| \le d$. Thus $f = (1 - n^{-1})w + n^{-1}e_j$ and

$$\begin{aligned}
&|\mathrm{Im}\ \phi(f)| \\
&\le |\mathrm{Im}\ \phi(w)| + |\phi(w) - \phi(f)| \\
&\le \alpha/2 + n^{-1} |\phi(e_j) - \phi(w)| \\
&\le \alpha/2 + n^{-1}(d + d) \\
&\le \alpha.
\end{aligned}$$

This completes the proof of Lemma 4.5.

4.6 REMARKS ON THE PROOF OF THEOREM 3.1

In the proof at certain stages simpler choices are indicated for the sequences constructed that provide a proof of all but a couple of the properties. At a first reading of the proof it is easiest to take the simplest choice at each stage. We shall use Lemmas 4.2 and 4.4 for right Banach A-modules as well as for left ones. The right versions may be deduced from the left ones by reversing the products and turning right mo ules over the algebra into left ones over the reversed algebra.

4.7 PROOF OF THEOREM 3.1

To prove properties 13 and 14 we require a suitable increasin sequence (γ_n) of positive real numbers, which will probably be tending infinity very fast. If one is not interested in properties 13 and 14, the the choice $\gamma_n = n$ gives all the other properties. Using the hypothesis that $\alpha_t \to \infty$ as $t \to \infty$, we choose an increasing sequence of real number (γ_n) such that

(1) $\gamma_o = 0$, $\gamma_n \ge \max (C,n)$, and $1 + \exp 2n \le \alpha_t$ for all $t \ge \gamma_n$. (

Here C is the constant of Property 14. We suppose that $||x|| \le 1$ (and $||y|| \le 1$).

Let $\{g_n : n \in \mathbb{N}\}$ be a countable bounded approximate identit for A. We shall be choosing a sequence (e_n) from a bounded approximat identity for A, but probably not from the sequence (g_n). By induction we choose a sequence (e_n) from a bounded approximate identity Λ for such that

(2) $||e_n|| \le 1$,

(3) $e_n \in \Lambda$,

(4) $|\mathrm{Im}\,\sigma(e_n)| \le 2^{-n-1}.\pi$ provided A is commutative, and

$\|(e_n - 1)(e_1+\ldots+e_{n-1})\|.\|w\| + \|(e_1+\ldots+e_{n-1})(e_n-1)\|.\|w\|$

$+ \|(e_n - 1).w\|$

and

$\|(e_n - 1)(e_1+\ldots+e_{n-1})\|.\|u\| + \|(e_1+\ldots+e_{n-1})(e_n - 1)\|.\|u\|$

$+ \|u.(e_n - 1)\|$

are so small that if $b_o^{\,t} = 1$ and $b_m^{\,t} = \exp t\left(\sum_{j=1}^{m} (e_j - 1)\right)$ for all $t \in \mathbb{C}$ and all $m \in \mathbb{N}$,

then

(5) $\|b_{n-1}^{\,t} - b_n^{\,t}\| \le \exp(-(n-1)\mathrm{Re}\ t).\ \{\exp(2|t|) - 1 + 2^{-n-1}\}$,

and

(6) $\|b_{n-1}^{\,t}.w - b_n^{\,t}.w\| \le 2^{-n}\ \delta$ and

$\|u\ .b_{n-1}^{\,t} - u.b_n^{\,t}\| \le 2^{-n}\ \delta$

for all $t \in \mathbb{C}$ with $|t| \le \gamma_n$, and all $w \in \{x,g_1,\ldots,g_n\}$ and all $u \in \{y,g_1,\ldots,g_n\}$.

Note that in (2) we have taken $\|e_n\| \le 1 \le d$ but that the choice of the sequence and the first few calculations just require $\|e_n\| \le d$. In (4) and (6) one of the norms is evaluated in a Banach module and the others are calculated in A. We shall use $e_n \in \Lambda$ in the proof of Property 9 (of Theorem 3.1), and we shall use inequality (4) in the proof of Property 7 and hence of Property 8.

We shall now choose $e_1, e_2, \ldots$ by induction so that (2) to (6) are satisfied. By Lemma 4.4(c) applied with $f = 0$ and $\mu = 0$, there is a constant ξ such that $\|w - b_1^{\,t}.w\| \le \xi\|(e - 1).w\|$ for all $t \in \mathbb{C}$ with $|t| \le \gamma_1$ and all $e \in \Lambda$ with $\|e\| \le d$. Then, by Lemma 4.5, we choose $e_1 \in \Lambda$ so that inequalities (4) and (6) are satisfied. We require Λ to be convex to apply 4.5. Recall that $\|\Lambda\| \le 1$ by the hypothesis of Property 9 so (2) holds. If we are not using Λ we must ensure that $\|e_1\| \le 1$. The choice of e_1 is possible because Λ is a convex bounded approximate identity for A bounded by 1, and the inequalities for $w = x$ and $u = y$ are handled using Lemma 4.2. Suppose that $e_1,\ldots,e_n$ have been chosen to satisfy (2) to (6). We choose e_{n+1} by applying Lemma 4.4 with $f = e_1+\ldots+e_n$ and $\mu = -n$, and using Lemma 4.5 to take care of inequality (4). The choice of $e_{n+1} \in \Lambda$ such that (2) to (6) hold is possible because Λ is a convex bounded approximate identity bounded by 1. Here we are again using Lemma 4.2 to ensure that $\|(e_{n+1} - 1).x\|$ and $\|y.(e_{n+1} - 1)\|$ can be chosen to be very small. This completes the inductive choice of the sequence (e_n).

We now define $t \mapsto a^t : H \to A$ and check that it is an analytic semigroup. For each $\nu > 0$, inequality (5) implies that $(b_n^{\,t})$ is uniformly Cauchy for t in the set $\{z \in \mathbb{C} : \operatorname{Re} z \geq \nu^{-1}, |z| \leq \nu\}$. This is because we have

$$\|b_n^{\,t} - b_{n-1}^{\,t}\| \leq \exp 2\nu.\exp(-(n - 1) \operatorname{Re} t) \leq \exp 2\nu.\exp(-(n - 1)\nu^{-1})$$

for $\gamma_n \geq \nu$ and $t \in \mathbb{C}$ with $\operatorname{Re} t \geq \nu^{-1}$ and $|t| \leq \nu$. Hence $\lim_n b_n^{\,t}$ exists in $A^{\#}$ for all $t \in H = \bigcup_{\nu>0} \{z \in \mathbb{C} : \operatorname{Re} z \geq \nu^{-1}, |z| \leq \nu\}$. Writing out the power series for $b_n^{\,t}$, we see that $b_n^{\,t} \in \exp(-nt) \oplus A$ and thus $\lim_n b_n^{\,t} \in A$ for all $t \in H$. We denote this limit by a^t. Each function $t \mapsto b_n^{\,t} : H \to A$ is a semigroup homomorphism from $(H,+)$ into $(A^{\#},\cdot)$ so $t \mapsto a^t : H \to A$ is a semigroup homomorphism. Also each function $t \mapsto b_n^{\,t} : H \to A^{\#}$ is analytic, and the sequence $(b_n^{\,t})$ converges uniformly to a^t on the compact set $\{z \in \mathbb{C} : \operatorname{Re} z \geq \nu^{-1}, |z| \leq \nu\}$. Therefore $t \mapsto a^t : H \to A$ is analytic.

We shall now define $t \mapsto x_t : \mathbb{C} \to X$, and we note that there is a similar definition of $t \mapsto y_t : \mathbb{C} \to Y$. By inequality (6) the sequence $(b_n^{\,-t}.x)$ is uniformly Cauchy for t in $\{z \in \mathbb{C} : |z| \leq \nu\}$ for each $\nu > 0$. Let $x_t = \lim_n b_n^{\,-t}.x$ for each $t \in \mathbb{C}$. Then $t \mapsto x_t : \mathbb{C} \to X$ is analytic. We can also define $t \mapsto G_{m,t}$ and $t \mapsto K_{m,t}$ from $\mathbb{C}$ into A, where

$$G_{m,t} = \lim_n b_n^{\,-t}.g_m, \quad \text{and}$$

$$K_{m,t} = \lim_n g_m.b_n^{\,-t}$$

for all $t \in \mathbb{C}$ and $m \in \mathbb{N}$.

With the construction of a^t, x_t and y_t complete, we check the various properties of Theorem 3.1.

Property 1. This property follows from the equality $x = b_n^t.(b_n^{-t}.x)$ for $t \in H$ and $n \in \mathbb{N}$, and from the definitions of a^t and x_t.

Property 2. Because $\{g_n : n \in \mathbb{N}\}$ is a bounded approximate identity for A, it is sufficient to show that $g_n \in (a^t.A)^-$ for all $n \in \mathbb{N}$ and $t \in H$ since $A = (\cup g_n A)^-$. The equality equivalent to $x = a^t.x_t$ (Property 1) for g_n is $g_n = a^t.G_{n,t}$ so $g_n \in a^t A$ for all $t \in H$.

Property 3. If $t \in H$ with $|t| \le \gamma_{m+1}$, then by inequality (5)

$$\begin{aligned} &\|a^t - b_m^t\| \\ &\le \sum_{k=m}^{\infty} \|b_{k+1}^t - b_k^t\| \\ &\le (\exp(2|t|) - 1) \sum_{k=m}^{\infty} \exp(-k \operatorname{Re} t) + \sum_{k=m}^{\infty} 2^{-k-2} \\ &\le (\exp(2|t|) - 1)(1 - \exp - \operatorname{Re} t)^{-1} + 1. \end{aligned}$$

Since $\gamma_1 \ge 1$ and $b_o^t = 1$ for all $t \in \mathbb{C}$, we have

$$\|a^t\| \le 2 + (\exp(2|t|) - 1)(1 - \exp(-\operatorname{Re} t))^{-1}$$

for all $t \in H$ with $|t| \le 1$. If

$$t \in U(\psi) = \{z \in \mathbb{C} : 0 < |z| \le 1, |\operatorname{Arg} z| \le \psi\},$$

then $\operatorname{Re} t \ge |t| \cos \psi$ so that

$$\begin{aligned} \|a^t\| &\le 2 + (\exp 2|t| - 1)(1 - \exp(-|t| \cos \psi))^{-1} \\ &\to 2 + \frac{2}{\cos \psi} \qquad \text{as } |t| \to 0. \end{aligned}$$

Hence $\{\|a^t\| : t \in U(\psi)\}$ is bounded.

Property 4. If $b \in A$ and $\varepsilon > 0$, then by Property 2 there is a $c \in A$ with $||b - a^1.c|| \le \varepsilon\, 2^{-1}(1 + \sup\{||a^t|| : t \in U(\psi)\})^{-1}$. We are using Property 3 to ensure that the supremum is finite. Thus

$$\begin{aligned} &||a^t.b - b|| \\ &\le (||a^t|| + 1)||b - a^1c|| + ||a^{t+1} - a^1||.||c|| \\ &< \varepsilon/2 + \varepsilon/2 \end{aligned}$$

for $t \in U(\psi)$ with $|t|$ sufficiently small.

Property 5. If $t > 0$, then

$$\begin{aligned} ||b_n{}^t|| &\le \exp(-nt) . \exp(t \sum_1^n ||e_j||) \\ &\le \exp(-nt).\exp(nt) = 1 \end{aligned}$$

for all $n \in \mathbb{N}$. Thus $||a^t|| \le 1$ for all $t > 0$.

Property 6. I do not know how to prove this property by the exponential methods that we are using here. We shall see how it can be deduced from the factorization theorem using the analytic functional calculus in Sinclair [1979a]. However we must then check that these functional calculus methods will give all the other properties of Theorem 3.1. The functional calculus methods require a commutative Banach algebra with a bounded approximate identity, and we obtain this commutative algebra by an application of Theorem 3.1 to A. By Theorem 3.1 with all the properties except 6 and 15 we obtain an analytic semigroup $t \mapsto a^t : H \to A$ (of course we must not use Property 6 to prove 7 to 14). We let B be the Banach subalgebra of A generated by the set $\{a^t : t > 0\}$. Then $(B.A)^- = A = (A.B)^-$ by Property 4, $x \in B.X$, $y \in Y.B$, and $B \cap \Lambda$ is a bounded approximate identity for B by Properties 3, 4, and 9. We shall apply Theorem 1 of Sinclair [1979a] to B and the left and right Banach B-modules X and Y. An examination of the statement and proof of Theorem 1 of Sinclair [1979a] will show that many of our properties are already included in that theorem either explicitly or implicitly in the proof. We are taking $a^t = 2^t\,\theta(w^t)$ for all $t \in H$, where θ and w are as in Theorem 1 of Sinclair [1979a]. From the working after inequality (8) of (iv) of that Theorem, for $0 < \alpha < x$ and $y \in \mathbb{R}$, we have

$$\begin{aligned}
&\|a^{x+iy}\| \\
&= 2^x \|\theta(w^{x+iy})\| \\
&\le \|w^{x+iy}\|_\alpha \,.\, 2^x.\{\varepsilon/2 + 2^\alpha.(2K - 1)(1 - K^{-\alpha})^{-1}\}.
\end{aligned}$$

Here $\varepsilon > 0$ and $K > 1$ are constants chosen at the beginning, and

$$\|w^{x+iy}\|_\alpha = \sup\{|z^{x+iy}|.|z|^{-\alpha} : z \in \mathbb{C},\ |z - 1/2| < 1/2\}.$$

We take $K = 2$ and using the maximum modulus theorem write the supremum in terms of z on the boundary

$$\{z \in \mathbb{C} : z = r\, e^{i\psi},\ r = \cos\psi,\ -\pi/2 \le \psi \le \pi/2\}$$

of the open disc $\{z \in \mathbb{C} : |z - 1/2| < 1/2\}$. Then

$$\begin{aligned}
\|a^{x+iy}\| &\le c_\alpha.2^x \,.\, \sup\{(\cos\psi)^{x-\alpha}.e^{-\psi y} : -\pi/2 \le \psi \le \pi/2\} \\
&\le c_\alpha.2^x.\left(1 + \left(\frac{y}{x-\alpha}\right)^2\right)^{-(x-\alpha)/2}.\exp\left(y \arctan\left(\frac{y}{x-\alpha}\right)\right) \\
&\le c_\alpha.2^x.\left(1 + \left(\frac{y}{x-\alpha}\right)^2\right)^{-(x-\alpha)/2}.\ \exp(\pi|y|/2).
\end{aligned}$$

The most straightforward inequality would be $\|a^{x+iy}\| \le c_\alpha.2^x \exp(\pi|y|/2)$, and even this I cannot obtain by exponential methods.

Property 7. This property requires an initial application of Theorem 3.1 except for properties 7 and 8 so we can pass from A to a commutative subalgebra B of A. The reason for this is that our spectral calculations require an application of Lemma 4.5, which has the hypothesis that the algebra is commutative. By Theorem 3.1 with properties 1 to 5 and 9 to 14 we choose an analytic semigroup $t \mapsto a^t : H \to A$. Note that we must be careful not to use 7 or 8 in the proof of the other properties. Let B be the Banach subalgebra of A generated by $\{a^t : t > 0\}$. Then $x \in (B.X)^-$, $y \in (Y.B)^-$, $(A.B)^- = A = (B.A)^-$, and $\Lambda \cap B$ satisfies the same hypotheses in B as Λ does in A. We now choose the sequence (e_n) in B to satisfy (2) to (6), and proceed as before working in $B^\#$ instead of $A^\#$. Let ϕ be a character on $B^\#$. If $t > 0$, then

$$
\begin{aligned}
&|\operatorname{Arg} \phi(b_n{}^t)| \\
&= |\operatorname{Arg} \exp (t \sum_1^n (\phi(e_j) - 1))| \\
&= t \, |\sum_1^n \operatorname{Im} \phi(e_j)| \\
&\le t \sum_1^n \pi . 2^{-j-1} \\
&\le \pi t/2 .
\end{aligned}
$$

Thus $|\operatorname{Arg} \phi(a^t)| \le \pi t/2$ for all $t > 0$ and all characters ϕ on B. Since the boundary of the spectrum of a^t in B contains the boundary of the spectrum of a^t in A, $\sigma(a^t) \subseteq U(\pi t/2)^-$ for all $t > 0$. Using the same idea we obtain $|\phi(a^t)| \le \exp(\pi \operatorname{Im} t/2)$ for all $t \in H$. This proves property 7.

Property 8. As in the proofs of Properties 6 and 7 we pass to the commutati Banach subalgebra B generated by $\{a^t : t > 0\}$. If B is a radical Bana algebra, then we can choose any semigroup $t \mapsto a^t : H \to B$ satisfying all of the properties of Theorem 3.1 except for 8, and it will then satisfy Property 8 by Theorem 3.6. Thus we may suppose that the carrier space Φ of B is not empty. We consider separately the two cases Φ compact and Φ locally compact.

Assume Φ is compact. Then there is an idempotent $e \in B$ such that $\phi(e) = 1$ for all $\phi \in \Phi$ (see Rickart [1960], Theorem 3.6.3 or Bonsa and Duncan [1973], Theorem 21.5). Now $t \mapsto (1 - e)a^t : (0,\infty) \to (1 - e)B$ is a continuous semigroup bounded by $(1 + \|e\|)$ into the radical Banach algebra $(1 - e)B$. Note that $(1 - e)B$ is non-zero because if $(1 - e)B$ were zero, then B would have an identity e so A would have an identit since $(A.B)^- = A = (B.A)^-$, contrary to hypothesis. Further $((1 - e)a^t (1 - e)B)^- = (1 - e)B$ for all $t > 0$. If $f \in L^1(\mathbb{R}^+)$ with $\theta(f) = 0$, then $(1 - e)\,\theta(f) = \int_0^\infty f(t)\,(1 - e)a^t\,dt = 0$ so $f = 0$ by Theorem 3.6 because $(1 - e)B$ is a non-zero radical Banach algebra.

Now assume that Φ is not compact. Let $\{h_n : n \in \mathbb{N}\}$ be a countable bounded approximate identity in B, and let $W_n = \{\phi \in \Phi : |h_n{}^\wedge(\phi)| > 1/2\}$ for each $n \in \mathbb{N}$. Then each W_n is an open subset of Φ with compact closure, and $\bigcup_n W_n = \Phi$ because $\{h_n : n \in \mathbb{N}\}$ is an approximate identity in B. Therefore Φ is σ-compact. We now choose a sequence (ϕ_n) in Φ such that $\phi_n(b) \to 0$ as $n \to \infty$ for all $b \in B$. For example, we could take $\phi_n \in W_{n+1} \setminus \bigcup_1^n W_j^-$ adjusting for the fact that $W_{n+1} \setminus \bigcup_1^n W_j^-$ may often be empty but is infinitely often non-empty; if $b \in B$ and $\varepsilon > 0$, then $\{\phi \in \Phi : |\phi(b)| \geq \varepsilon\}$ is compact so is contained in $\bigcup_1^n W_j$ for some n, and hence $|\phi_k(b)| < \varepsilon$ for all $k > n$. We now choose a continuous function h on Φ vanishing at infinity and satisfying $h(\phi_n) \geq 1/n$ for all $n \in \mathbb{N}$ - we are assuming that the points in the sequence (ϕ_n) are distinct. We regard the Banach space $C_o(\Phi)$ of continuous functions vanishing at infinity on Φ as a Banach B-module by defining $a.f = \hat{a}f$ for all $a \in B$ and $f \in C_o(\Phi)$, where $\wedge$ is the Gelfand map from B into $C_o(\Phi)$. Since B has a bounded approximate identity, $B.C_o(\Phi)$ is dense in $C_o(\Phi)$. We apply Theorem 3.1 with all the properties except 8 to the Banach algebra B, the left Banach B-module $C_o(\Phi) \oplus X$, the right Banach B-module Y, and the elements $h \oplus x$ and y in $C_o(\Phi) \oplus X$ and Y. Then there is an analytic semigroup $t \mapsto b^t : H \to B$ satisfying Theorem 3.1 except for 8, and an entire function $t \mapsto h_t : \mathbb{C} \to C_o(\Phi)$ such that $h = b^t.h_t$ for all $t \in H$. Further $\sigma(b^t) \subseteq U(r\pi/2)^-$ for all $r > 0$. The semigroup $t \mapsto b^t$ is the required semigroup. For each $n \in \mathbb{N}$, the map $t \mapsto \phi_n(b^t) : H \to \mathbb{C}$ is a non-zero analytic semigroup, and so there is a $\beta_n \in \mathbb{C}$ such that $\phi_n(b^t) = \exp(-t\beta_n)$ for all $t \in H$. Since $\|b^t\| \leq 1$ for $t > 0$, Re $\beta_n \geq 0$ for all n. Since $b_1{}^\wedge \in C_o(\Phi)$, $\phi_n(b^1) = \exp(-\beta_n)$ tends to zero as n tends to infinity. Thus Re β_n tends to infinity as n tends to infinity. Also

$$|\exp(-\beta_n)| \geq |h(\phi_n)| \, \|h_1\|^{-1} \geq \|h_1\|^{-1} n^{-1},$$

since $h(\phi_n) = \phi_n(b^1) h_1(\phi_n)$, so that Re $\beta_n \leq \log n + \log \|h_1\|$ for all n. Hence the series $\sum 1/\text{Re}\, \beta_n$ summed over n with Re $\beta_n \neq 0$ diverges. Let $f \in L^1(\mathbb{R}^+)$ with $\theta(f) = 0$, and let $\mathcal{L}f$ denote the Laplace transform of f. Then

$$(Lf)(\beta_n) = \int_0^\infty f(t)\exp(-\beta_n t)\,dt$$

$$= \phi_n\left(\int_0^\infty f(t)\, b^t\, dt\right)$$

$$= \phi_n \quad \theta(f) = 0$$

for all n. By Corollary A 1.4 it follows that Lf (and so f) is zero. This proves property 8.

Property 9. The sequence (e_n) was chosen in (3) to be in Λ. The convexity of Λ implies that $f_n = n^{-1}\sum_1^n e_j \in \Lambda$, and the closure of Λ under powers ensures that $f_n{}^j \in \Lambda$ for all $j, n \in \mathbb{N}$.
Thus

$$b_n{}^t = \exp\left(t\sum_1^n (e_j - 1)\right)$$

$$= \exp(-tn)\exp(tnf_n)$$

$$= \exp(-tn)\sum_{j=0}^{\infty}\frac{(tn)^j}{j!} f_n{}^j$$

$$\in \exp(-tn)\{1 + (\exp(tn) - 1)\Lambda\}$$

for all $t > 0$ and all $n \in \mathbb{N}$. Taking the limit as $n \to \infty$, we have $a^t \in \Lambda$ for all $t > 0$.

Property 10. This property follows directly from the definitions of a^t and x_t.

Property 11. If $t \in H$ and $z \in \mathbb{C}$, then $b_n{}^{-z}.x = b_n{}^t.(b_n{}^{-z-t} x)$ for all n, and taking limits as n tends to infinity gives the result.

Property 12. Lemma 4.2 and the hypothesis concerning x ensure that $x \in (A.x)^-$. Thus $b_n{}^t x \in \exp nt.\,(x + A.x) \subseteq (A.x)^-$ for all $t \in \mathbb{C}$ and all $n \in \mathbb{N}$. Therefore $x_t \in (A.x)^-$ for all $t \in \mathbb{C}$.

Property 13. If $t \in \mathbb{C}$ with $\gamma_m < |t| \le \gamma_{m+1}$, then

$$||x_t|| \le ||b_m^{-t}.x|| + \sum_{k=m+1}^{\infty} ||b_{k-1}^{-t}.x - b_k^{-t}.x||$$

$$\le \exp(2m|t|) + \sum_{k=m+1}^{\infty} 2^{-k}$$

by the definition of b_m^{-t}, the normalization $||x|| \le 1$, and inequality (6). Thus

$$||x_t|| \le \exp(2m|t|) + 1 \le (\exp 2m + 1)^{|t|} \le \left(\alpha_{|t|}\right)^{|t|}.$$

by inequality (1). Thus $||x_t|| \le \left(\alpha_{|t|}\right)^{|t|}$ for all $t \in \mathbb{C}$ with $|t| \ge 1$. Note we have used the inequality $1 + \zeta^r \le (1 + \zeta)^r$ for $\zeta > 0$ and $r \ge 1$.

Property 14. If $t \in \mathbb{C}$ with $|t| \le C$, then $|t| \le \gamma_1$ and

$$||x - x_t|| \le \sum_1^{\infty} ||b_{k-1}^{-t}.x - b_k^{-t}.x||$$

$$\le \sum_1^{\infty} 2^{-k} \delta = \delta$$

by inequality (6).

Property 15. The only way I know of proving this property is by the analytic functional calculus methods of Sinclair [1979]. It is really a functional calculus result. Those methods work because the function $z \mapsto (z^\alpha - z^\beta)^t = \exp t(\log (z^\alpha - z^\beta))$ is analytic on the open disc $\{z \in \mathbb{C} : |z - 1/2| < 1/2\}$, and is dominated by $|z|^\zeta$ for some $\zeta > 0$ as $|z| \to 0$ with z in the disc. Thus the function is in the algebra ζ of Theorem 1 of Sinclair [1979].

This completes the proof of Theorem 3.1 and we turn to the proof of Theorem 3.15.

4.8 PROOF OF THEOREM 3.15

In proving the three properties of this result we have an extra condition to impose on the choice of the sequence (e_n) - a condition ensuring that (e_n) is nice with respect to the derivation, multiplier, or automorphisms. We shall deduce Property 17 from 16.

<u>Property</u> 16. Let $\{D_n : n \in \mathbb{N}\}$ be a countable dense subset of the unit ball of the separable Banach space Z of derivations on A. In making the initial choice of the sequence (e_n) at the beginning of 4.7 we require the additional inequality

$$(7) \quad \|D_j(e_n)\| \le 2^{-n-1} e^{-1} \quad \text{for } 1 \le j \le n,$$

and the stronger version of (5)

$$(5') \quad \|b_n^{\,t} - b_{n-1}^{\,t} + \exp(-(n-1)t) - \exp t(e_n - n)\|$$
$$\le \exp(-(n-1) \operatorname{Re} t) . \; 2^{-n-1}$$

for all $t \in \mathbb{C}$ with $|t| \le n$ and all $n \in \mathbb{N}$.

The inductive choice of the sequence (e_n) proceeds exactly as in 4.7 with the additional inequality (7) built into the choice. We are able to ensure that (7) is satisfied because of the hypotheses on Λ and Z. We obtain (5´) from Lemma 4.4(a) with $f = e_1 + \ldots + e_{n-1}$ and $e = e_n$ by choosing $\|(e - 1)f\|$ and $\|f(1 - e)\|$ very small. From Lemma 4.4(a).

$$\|b_n^{\,t} - b_{n-1}^{\,t} + \exp -(n-1)t - \exp t(e_n - n)\|$$
$$= \exp(-(n-1) \operatorname{Re} t) . \|\exp t(f + e_n - 1) - \exp tf + 1 - \exp t(e_n - 1)\|$$
$$\le \exp(-(n-1) \operatorname{Re} t) . (\exp(n|t|) - 1) . \{\|(e_n - 1)f\| + \|f(e_n - 1)\|\},$$

which gives (5´) on choosing e_n so that $\|(e_n - 1)f\|$ and $\|f(e_n - 1)\|$ are small enough. This completes the inductive choice of the sequence (e_n).

The derivations D in Z are lifted to $A^{\#}$ by defining $D(1) = 0$. Using the formula for the action of a derivation on the kth-power of an element we obtain

$$\|D(\exp t\, e_n)\|$$
$$\le \sum_{k=1}^{\infty} \frac{|t|^k}{k!} k \|e_n\|^{k-1} \|D(e_n)\|$$
$$\le |t| \, \|D(e_n)\| \exp |t|$$

for all $t \in \mathbb{C}$ and all $D \in Z$.

From inequality (5´) it follows that

$$\|D_j(b_n^t) - D_j(b_{n-1}^t)\|$$

$$\leq \exp(-(n-1)\operatorname{Re} t).\,2^{-n-1} + \exp(-n \operatorname{Re} t).\,\|D_j(\exp t\, e_n)\|$$

$$\leq \exp(-(n-1)t).2^{-n}$$

for $0 < t \leq 1$ and $1 \leq j \leq n$. Thus if $0 < t \leq 1$ and $j < m \in \mathbb{N}$, we have

$$\|D_j(a^t)\|$$

$$\leq \|D_j(b_m^t)\| + \sum_{n=m+1}^{\infty} \|D_j(b_n^t) - D_j(b_{n-1}^t)\|$$

$$\leq \|D_j(b_m^t - 1)\| + \sum_{n=m+1}^{\infty} \exp(-(n-1)t).2^{-n}$$

$$\leq \|b_m^t - 1\| + 2^{-m}$$

$$\leq \exp(2mt) - 1 + 2^{-m}$$

Let $\varepsilon > 0$. Then there is an $m > j$ such that $2^{-m} < \varepsilon/2$, and corresponding to this m there is $\nu > 0$ such that $\exp(2mt) - 1 < \varepsilon/2$ for $0 < t < \nu$. Therefore $\|D_j(a^t)\| \leq \varepsilon$ for $0 < t < \nu$. Since $\{D_j : j \in \mathbb{N}\}$ is dense in the unit sphere of Z and since $\{\|a^t\| : 0 < t < 1\}$ is bounded, the proof of Property 16 is complete.

<u>Property</u> 17. We shall deduce this from Property 16. Let $\{c_j : j \in \mathbb{N}\}$ be a countable subset in B such that the set $\{c_j + a : j \in \mathbb{N}\}$ is dense in B/A. Let Z be the closure in $BL(A)$ of the linear span of the set of derivations $\operatorname{ad}(c_j)$, where $\operatorname{ad}(c_j).a = c_j a - ac_j$ for all $a \in A$. The existence of a quasicentral bounded approximate identity in Λ for B implies that there is a bounded approximate identity (g_n) in Λ such that $\operatorname{ad}(c_j)g_n \to 0$ as $n \to \infty$ for all $j \in \mathbb{N}$. From Theorem 3.1 and 3.15, Property 16, we obtain an analytic semigroup $t \mapsto a^t : H \to A$ such that $\|D(a^t)\| \to 0$ as $t \to 0$, $t > 0$ for all $D \in Z$. Finally let E be the set of all continuous linear operators T on A satisfying $T(a^t) \to 0$ as $t \to 0$. Because the set $\{\|a^t\| : 0 < t \leq 1\}$ is bounded, E is a Banach subalgebra of $BL(A)$ that does not contain the identity operator. From

the construction of $t \mapsto a^t$ the space of derivations Z is contained in E and $ad(a) \in E$ for all $a \in A$ (by 3.1, Property 4). This proves 17.

Property 18. The proof of this is similar to the proof of property 16 - we give some of the details. We require inequality (5´) as in the proof of 16. Let $\{\beta_n : n \in \mathbb{N}\}$ be a countable dense subset of G. Using the hypothesis on the approximate identity Λ we may ensure that the sequence $(e_n) \subseteq \Lambda$ satisfies

$$(8) \quad \|\beta_j(e_n) - e_n\| \leq 2^{-n} \text{ for } 1 \leq j \leq n.$$

The automorphisms ϕ on A are lifted to $A^{\#}$ by defining $\phi(1) = 1$.

For each $n \in \mathbb{N}$, $\beta \in G$, and $0 < t$, we have

$$\|\beta(b_n^t) - \beta(b_{n-1}^t) - b_n^t + b_{n-1}^t\|$$
$$\leq (\|\beta\| + 1)\ \|b_n^t - b_{n-1}^t + \exp(-(n-1)t) - \exp t(e_n - n)\|$$
$$+ \|(\beta - 1) \exp t(e_n - n)\|.$$

The last term is bounded above by

$$\exp(-tn) \sum_1^\infty \frac{t^k}{k!} \|\beta(e_n)^k - e_n^k\|$$
$$\leq \exp(-tn) \sum_1^\infty \frac{t^k}{k!} \|\beta(e_n) - e_n\| . \|e_n\|^{k-1} (\|\beta\| + 1)^k$$
$$\leq \exp(-tn) . \{\exp(t(\|\beta\| + 1)) - 1\} \|\beta(e_n) - e_n\| .$$

Combining these two inequalities, (8), and (5´), we have

$$\|\beta_j(b_n^t) - \beta_j(b_{n-1}^t) - b_n^t + b_{n-1}^t\|$$
$$\leq (\|\beta_j\| + 1) \exp(-(n-1)t) . 2^{-n-1} +$$
$$\exp(-tn) \{\exp(t(\|\beta_j\| + 1)) - 1\} . 2^{-n}$$

for $0 < t < 1$ and $1 \leq j \leq n$. If $0 < t < 1$ and $j \leq m$, then

$$\|\beta_j(a^t) - a^t\|$$

$$\leq \|\beta_j(b_m{}^t) - b_m{}^t\| + \sum_{m+1}^{\infty} \|\beta_j(b_n{}^t) - \beta_j(b_{n-1}{}^t) - b_n{}^t + b_{n-1}{}^t\|$$

$$\leq (\|\beta_j\| + 1)\ \{\exp\ (2mt) - 1\} + 2^{-m}\ (\|\beta_j\| + 1 + \exp\ (t\|\beta_j\| + 1) - 1)$$

by working similar to that used in 16. For given $\varepsilon > 0$ we choose m very large, and then choose t small depending on m. The details are similar to those in 16. This completes the proof of Theorem 3.15.

4.9 NOTES AND REMARKS ON CHAPTER 4.

Doran and Wichman [1979] discuss Cohen's factorization theorem in great detail in their lecture notes on bounded approximate identities and factorization. There is a historical discussion in those notes and in Hewitt and Ross [1970]. Lemma 4.2 is what permits one to show that the subset $A.X$ of the Banach A-module X is actually a closed submodule. This step was found independently by Hewitt [1964], Curtis and Figá-Talamanca [1966], and Gulick, Liu, and van Rooij [1967]. Lemmas 4.3 and 4.4 are minor modifications of Lemmas in Sinclair [1978]. Lemma 4.5 is a special case of Theorem 2.2 of Dixon [1980] that is just strong enough for our purposes.

Our proof of Theorem 3.1 is a variant of the proof of Theorem 1 of Sinclair [1978] except that properties 6 and 15 are deduced from Theorem 1 of Sinclair [1979]. In the proof of Theorem 3.1 we approximate not only the single element x but also a countable bounded approximate identity (g_n). This is the technique used in Varopoulos [1964] and Johnson [1966]. Theorem 3.15(a) and (c) are new but essentially like 3.15(b). Theorem 3.15(b) is proved in Sinclair [1979a] by analytic functional calculus methods.

5 RESTRICTIONS ON THE GROWTH OF $\|a^t\|$

5.1 INTRODUCTION

Let $t \mapsto a^t : H \to A$ be an analytic semigroup from the open right half plane H into a Banach algebra A such that $(a^t A)^- = A$ for all $t \in H$. In this Chapter we are concerned with relationships between the structure of the Banach algebra A and the growth of $\|a^t\|$ as t approaches the boundary of H in some way. We consider three types of approach to the boundary of H : along a ray in H emanating from 0, along vertical lines in H, and in a semidisc centred at 0 in H. Each type is discussed in a section of this Chapter. The more complicated boundary of H compared with $\mathbb{R}^+$ means that it is possible to extract more structure from restrictions on an analytic semigroup defined on H than a continuous semigroup defined on $[0,\infty)$.

In Chapter 3 we found that in each radical Banach algebra A with a bounded approximate identity there are analytic semigroups $t \mapsto a^t$ such that $\|a^{re^{i\theta}}\|^{1/r}$ tends to zero arbitrarily slowly as r tends to infinity for $|\theta| < \pi/2$. Rather surprisingly for analytic semigroups $\|a^{re^{i\theta}}\|^{1/r}$ cannot tend to zero arbitrarily fast as r tends to infinity. This is the conclusion of Theorem 5.3, and in Theorem 5.5 we shall show that the analytic property is crucial - without analyticity results of this type fail.

There is a link between the behaviour of $\|a^{1+iy}\|$ as $|y| \to \infty$ and the fine local behaviour of a^t for t near zero in H. If t is near zero in H, then $(\mathrm{Re}\ t)^{-1} t \in 1 + i\mathbb{R}$ so that a large power, $(\mathrm{Re}\ t)^{-1}$, of a^t moves a^t into $\{a^{1+iy} : y \in \mathbb{R}\}$. We can also see this intuitive idea by considering the fractional integral semigroup $t \mapsto I^t : H \to L^1(\mathbb{R}^+)$ into $L^1(\mathbb{R}^+)$. Much of the mass of I^{1+iy} occurs near zero in $\mathbb{R}^+$, and so in the region of $\mathbb{R}^+$ where approximate identities occur in $L^1(\mathbb{R}^+)$ though I^{1+iy} will have complex values as a function on $\mathbb{R}^+$ rather than positive values. To see that much of $|I^{1+iy}|$ is concentrated near zero we consider the interval $[0,1]$, but $[0,\varepsilon]$ would do just as well. Now $t \mapsto I^t|_{[0,1]} : H \to L^1_*[0,1]$ is an analytic semigroup because the restriction map is a natural quotient from $L^1(\mathbb{R}^+)$ onto the Volterra algebra $L^1_*[0,1]$. By Theorem 5.6

$$\int_{\mathbb{R}} (1 + y^2)^{-1} \log^+ \| I^{1+iy}|_{[0,1]} \|_1 \, dy$$

$$= \int_{\mathbb{R}} (1 + y^2)^{-1} \log^+ \{\int_0^1 |I^{1+iy}(w)| \, dw\} \, dy$$

is infinite. Thus much of the area under $w \mapsto I^{1+iy}(w) : \mathbb{R}^+ \to \mathbb{R}^+$ must be concentrated near $w = 0$ for large $|y|$. Of course we could calculate this directly for the fractional integral semigroup, but the general idea above applies to any analytic semigroup from H into $L^1(\mathbb{R}^+)$. In Theorem 5.6 we shall see that the finiteness of

$$\int_{\mathbb{R}} (1 + y^2)^{-1} \log^+ \|a^{1+iy}\| \, dy$$

for $t \mapsto a^t : H \to A$ an analytic semigroup with $(a^1A)^- = A$ implies that there are no continuous non-zero homomorphisms from A into a radical Banach algebra. Intuitively this is because the semigroup near zero looks so like an identity that it cannot occur in a radical Banach algebra.

In Theorem 5.14 we shall show that boundedness of the semigroup in the semidisc $\{z \in H : |z| \leq 1\}$ centred at 0 in H implies that the multiplier algebra is non-separable. An example shows that there are separable Banach algebras with bounded approximate identities (and hence analytic semigroups) with separable multiplier algebras. There are other restrictions on the semigroup that we have not investigated: for example, how does $\|a^t\| \leq 1$ and $(a^tA)^- = A$ for all $t \in H$ affect the structure of the algebra? Semigroups like this exist in $C_0(\mathbb{R})$.

5.2 GROWTH ON RAYS - LOWER RATES OF GROWTH IN RADICAL ALGEBRAS

In this section we shall discuss the growth of $\|a^t\|$ as t tends to infinity along a ray emanating from 0 in the open right half plane H. Firstly the rate of growth is always at most exponential along a ray because the boundedness of the set $\{\|a^t\| : t \in \text{ray}, 1 \leq |t| \leq 2\}$ may be pushed along the ray by the semigroup property to give an exponential bound of the form $C.M^{|t|}$. What about a lower bound? If there is a character ϕ on the Banach algebra such that $\phi(a^z) \neq 0$ for some $z \in H$, then there is a complex number ρ such that $\phi(a^t) = \exp(t\rho)$ for all $t \in H$. Now if $|\theta| < \pi/_2$, then $\|a^{re^{i\theta}}\| \geq |\phi(a^{re^{i\theta}})| = |\exp(\rho re^{i\theta})| = \exp$ for all $r > 0$ and some $k \in \mathbb{R}$. In this case the growth is exponential along each ray in H. However if the Banach algebra is a radical algebra we easily see that $\|a^{re^{i\theta}}\|^{1/r} \to 0$ as $r \to \infty$ by using the spectral radius formula. The rate of growth of $\|a^t\|$ along a ray is less than exponential. But how fast does $\|a^t\|$ tend to zero? Corollary 3.13 shows that in a radical Banach algebra A with a (countable) bounded approximate identity for a prescribed rate of decrease to zero there is an analytic semigroup $t \mapsto a^t : H \to A$ such that $\|a^t\|^{1/|t|}$ tends to zero more slowly than the specified rate as t tends to infinity in H. Thus analytic semigroups may be chosen in radical Banach algebras with countable bounded approximate identities such that $\|a^t\|^{1/|t|}$ tends to zero very slowly. Can they tend to zero very fast? The answer is no, and we prove this in Theorem 5.3.

Though $r^{-1} \log\|a^{re^{i\theta}}\|$ tends to $-\infty$ as r tends to infinity, the following theorem shows that this convergence to $-\infty$ is slower than $-r^{\alpha}$ for any $\alpha > 0$.

5.3 THEOREM

Let A be a radical Banach algebra, let X be a left Banach A-module, and let $t \mapsto a^t$ $H \to A$ be an analytic semigroup. If $\gamma > 1$ and $x \in X$ with $a^1.x \neq 0$, then $\lim_{r\to\infty} r^{-\alpha} \log\|a^{re^{i\psi}}.x\| = 0$ for all $\psi \in (-\pi/2, \pi/2)$. The convergence is uniform for $\Psi \in [-\alpha,\alpha]$ for all $\alpha \in (0,\pi/2)$.

Proof. Since $\|a^{re^{i\psi}}\| \to 0$ as $r \to \infty$, we have

$$(1) \quad \limsup_{r\to\infty} r^{-\gamma} \log\|a^{re^{i\psi}}.x\| \leq 0$$

for all $\psi \in (-\frac{\pi}{2},\frac{\pi}{2})$. We shall prove that the lim inf is non-negative, and except for the uniformity of the limit the result will follow from this.

Choose β such that $\gamma^{-1} < \beta < 1$. By the Hahn-Banach Theorem we choose $f \in X^*$ such that $f(a^1.x) \neq 0$, and we let $F(z) = f(a^{1+(1+z)^\beta}.x)$ for all z in a neighbourhood of the closed right half plane H^-, where z^β is the principal power of z. Note that if $z \in H^-$, then $(1 + z)^\beta \in H$, and that F is an analytic function on a neighbourhood of H^-. The 1+ in the exponent of a ensures the analyticity of F in a neighbourhood of H^-, and also gets around the problem that $t \mapsto a^t : H \to A$ may not be bounded near zero. Let $\alpha \in (0, \pi/2)$ and let

$$(2) \quad M = \sup \{\|a^z\| : 1/_2 \le |z| \le 2, \ |\text{Arg } z| \le \alpha\}$$

If $w \in H$ with $|\text{Arg } w| \le \alpha$ and $|w| \ge 1$, then there is a positive integer n such that $n \le |w| \le n + 1$. For this w and n we have $|\text{Arg}(w/n)| \le \alpha$, and $\|a^w\| \le \|a^{w/n}\|^n \le M^n \le M^{|w|}$. If z is in H^-, then taking $w = (1 + z)^\beta$, we obtain

$$|F(z)| \le \|f\|.\|x\| \, M^{1 + |1+z|^\beta}$$

Thus $F : H \to \mathbb{C}$ is of exponential type (that is, there are constants a and b such that $|F(z)| \le a \exp(b|z|)$ for all $z \in H$), and

$$\int_{\mathbb{R}} \frac{\log^+ |F(iy)|}{1 + y^2} \, dy$$
$$\le \int_{\mathbb{R}} \left\{ \frac{1 + (1 + y^2)^{\beta/2}}{1 + y^2} \cdot \log^+ M + \frac{1}{1+y^2} \log^+ (\|f\|.\|x\|) \right\} dy$$

is finite. By the Ahlfors-Heins Theorem (Theorem A1.1) there is a real number c such that $\lim_{r\to\infty} r^{-1} \log |F(re^{i\phi})| = c \cos \phi$ for almost all $\phi \in (-\pi/2, \pi/2)$. For $\phi \in (-\pi/2, \pi/2)$ we have the inequality

$$r^{-1} \log |F(re^{i\phi})|$$
$$\le r^{-1} \log \|f\| + r^{-1} \log \|a^{1+G(r)}\| + r^{-1} \log \|a^{(re^{i\phi})^\beta}.x\|$$

for all $r > 0$, where

$$G(r) = (1 + re^{i\phi})^\beta - (re^{i\phi})^\beta .$$

For large r we have $G(r) = r^{\beta} e^{i\phi\beta} \{(1 + r^{-1} e^{-i\phi})^{\beta} - 1\}$ is approximately $\beta r^{\beta-1} e^{i\phi(\beta - 1)}$ so $G(r)$ tends to zero as r tends to infinity. Therefore $\lim\inf r^{-1} \log \|a^{(re^{i\phi})}\| \geq c \cos\phi$ for almost all $\phi \in (-\pi/2, \pi/2)$, and so $\lim\inf r^{-1/\beta} \log \|a^{re^{i\psi}}.x\| \geq c \cos(\psi/\beta)$ for almost all $\psi \in (-\beta\pi/2, \beta\pi/2)$.

We shall now check the limit required in the conclusion for $\psi = 0$, and deduce the general case and the uniformity of the limit from this. We choose $\chi \in (0, \beta\pi/2)$ such that $\lim\inf r^{-1/\beta} \log \|a^{re^{i\chi}}.x\| \geq c \cos(\chi/\beta)$. From the following diagram we see that there are $\xi, \zeta > 0$ such that $\zeta e^{i\chi} = 1 + \xi e^{i\beta\pi/2}$.

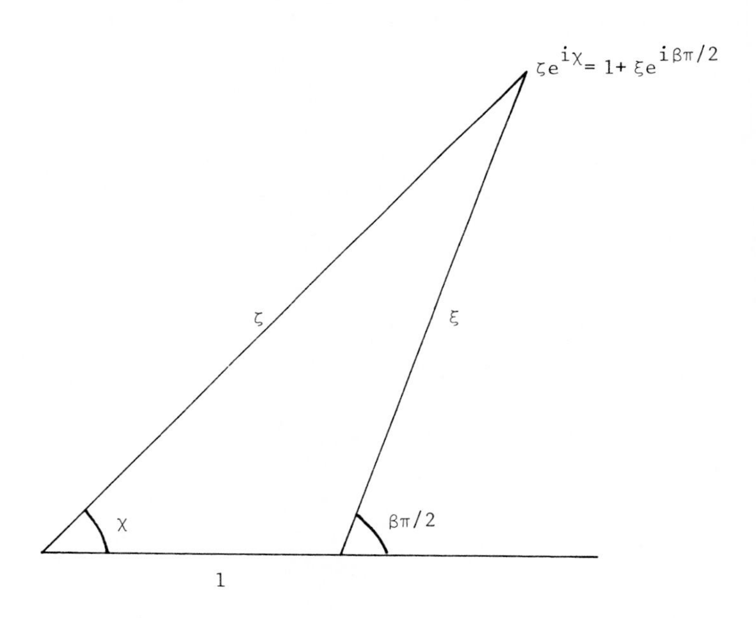

For large r we have $\|a^{r\xi e^{i\beta\pi/2}}\| \leq 1$, because the Banach algebra is a radical algebra, and hence $\log \|a^{\zeta re^{i\chi}}.x\| \leq \log \|a^{r}.x\|$. Thus

$$\lim\inf r^{-1/\beta} \log \|a^{r}.x\| \geq \zeta^{1/\beta} c \cos(\chi/\beta) \geq 0.$$

From this, $r^{-\gamma} \log \|a^{r}.x\| = (r^{-\gamma+1/\beta}).r^{-1/\beta} \log \|a^{r}.x\|$, and $-\gamma + 1/\beta < 0$

it follows that $\liminf r^{-\gamma} \log \|a^r.x\| \geq 0$. This inequality and (1) imply that $\liminf r^{-\gamma} \log \|a^r.x\| = 0$.

If $\psi \in [-\alpha, \alpha]$, then $\nu = e^{i\psi} + e^{-i\psi} \geq 2 \cos \alpha > 0$, and

$$\begin{aligned} & r^{-\gamma} \log \|a^{r\nu}.x\| \\ & \leq r^{-\gamma} \log \|a^{re^{i\psi}}.x\| + r^{-\gamma} \log \|a^{re^{-i\psi}}\| \\ & \leq r^{-\gamma} \log \|a^{re^{i\psi}}.x\| + r^{-\gamma}.r.\log M \end{aligned}$$

for all $r \geq 1$ by (2). Hence

$$r^{-\gamma} \log \|a^{re^{i\psi}}.x\| \geq \nu^{\gamma}(r\nu)^{-\gamma} \log \|a^{r\nu}.x\| - r^{-\gamma+1} \log M$$

and

$$r^{-\gamma} \log \|a^{re^{i\psi}}.x\| \leq r^{-\gamma}\, r \log M$$

for all $r \geq 1$ by (2). Taking the limit as r tends to infinity in these inequalities proves the result. The uniformity of the convergence for $\psi \in [-\alpha, \alpha]$ is obtained using $\nu \geq 2 \cos \alpha > 0$.

In Theorem 5.4 we shall give a construction that shows that the above result depends in a crucial way on analyticity. We shall construct a continuous semigroup on $(0,\infty)$ such that $\|a^t\|$ tends to zero very fast as t tends to infinity. Recall that a radical weight ω on $\mathbb{R}^+$ is a continuous function from $\mathbb{R}^+$ into $(0,\infty)$ such that $\omega(x+y) \leq \omega(x)\,\omega(y)$ for all $x,y \in \mathbb{R}^+$ and $\omega(t)^{1/t} \to 0$ as $t \to \infty$.

5.4 <u>THEOREM</u>

Let g be any continuous positive function on $[2,\infty)$. Then there is a radical weight ω on $\mathbb{R}^+$ and a continuous semigroup $t \mapsto a^t : (0,\infty) \to L^1(\mathbb{R}^+,\omega)$ such that $\|a^t\| \leq g(t)$ for all $t \geq 2$, and $a^t * b \to b$ as $t \to 0$, $t > 0$ for all $b \in A$.

<u>Proof</u>. Firstly we shall choose the weight ω by a constructive process from g with ω tending to zero very fast near infinity, and then we shall show that a natural semigroup in $L^1(\mathbb{R}^+) \subseteq L^1(\mathbb{R}^+,\omega)$ has the required properties. We define the continuous positive function ψ on $[0,\infty)$ by

$$\psi(x) = \begin{cases} 1 & \text{for } 0 \le x \le 1 \\ \text{linear between 1 and 2} \\ \inf(1, g(x)) & \text{for } x \ge 2 \end{cases}$$

We define $\phi : [0,\infty) \to \mathbb{R}^+$ by

$$\phi(x) = \inf \{\psi(x_1)\cdot\cdots\cdot\psi(x_n) : x_1,\cdots,x_n \ge 0,\ x = x_1+\cdots+x_n,\ n \in$$

From the definition of ϕ it follows that $0 \le \phi(x) \le 1$ for all $x \ge 0$, $\phi(x) = 1$ for $0 \le x \le 1$, $0 \le \phi(x) \le g(x)$ for all $x \ge 2$, and $\phi(u + w) \le \phi(u)\ \phi(w)$ for all $u,w \ge 0$. We shall now show that ϕ is positive and continuous.

Let $x \ge 1$, and let $\{x_1,\cdots,x_n\}$ be a finite set of positive real numbers such that $x = x_1+\cdots+x_n$. Let m denote the largest integer with $m \le x$. At most m elements of the set $\{x_1,\cdots,x_n\}$ are greater than 1, and those $x_j < 1$ give $\psi(x_j) = 1$ so that

$$\psi(x_1)\cdot\cdots\cdot\ \psi(x_n) \ge (\inf \{\psi(z) : 0 \le z \le x\})^m > 0$$

since ψ is a continuous positive function on $[0,\infty)$. Thus $\phi(x) \ge (\inf \{\psi(z) : 0 \le z \le x\})^m > 0$. If $y > x$ and if $x = x_1+\cdots+ x_n$ with $x_1,\cdots,x_n$ positive real numbers, then

$$\begin{aligned}\phi(y) &\le \psi(y - x)\cdot\ \psi(x_1)\cdot\cdots\cdot\ \psi(x_n) \\ &\le \psi(x_1)\cdot\cdots\cdot\ \psi(x_n).\end{aligned}$$

Taking the infimum over all such finite sets $\{x_1,\cdots,x_n\}$ we obtain $\phi(y) \le \phi(x)$. Thus ϕ is decreasing.

Let $k > 0$ and $\varepsilon > 0$. Now ψ is uniformly continuous on $[0,k]$ so there is a $\delta > 0$ with $\delta < 1$ and $|\psi(u) - \psi(w)| < \varepsilon/2$ for $u,w \in [0,k]$ and $|u - w| < \delta$. If $1 \le x < y \le k$ and if $y < x + \delta$, then $\phi(y) \le \phi(x)$. There are $y_1,\cdots,y_n \in [0,k]$ such that $y = y_1+\cdots+y_n$ and

$$\psi(y_1)\cdot\cdots\cdot\ \psi(y_n) < \phi(y) + \varepsilon/2.$$

Because $\psi(y_j) \le 1$ for all j and $\psi(y_1 + x - y) < \psi(y_1) + \varepsilon/2$, we have

$$\begin{aligned}\phi(x) &\le \psi(y_1 + x - y) \cdot \psi(y_2) \cdot \cdots \cdot \psi(y_n) \\ &\le \psi(y_1) \cdot \cdots \cdot \psi(y_n) + \varepsilon/2 \\ &< \phi(y) + \varepsilon.\end{aligned}$$

Hence ϕ is continuous on $[0,\infty)$.

Let $\omega(x) = \phi(x) \exp(-x^2)$ for all $x \ge 0$. Then ω is a weight on $\mathbb{R}^+$ since ϕ and $\exp(-x^2)$ are weights, and is a radical weight since $\exp(-x^2)$ is a radical weight and ϕ is bounded. Further $\omega(x) \le g(x)$ for all $x \ge 2$.

The injective map $f \mapsto f : L^1(\mathbb{R}^+) \to L^1(\mathbb{R}^+,\omega)$ is a continuous monomorphism of norm 1. Let δ^t be the unit point mass at t in $[0,\infty)$. The semigroup $t \mapsto \delta^t : \mathbb{R}^+ \to M(\mathbb{R}^+) = \mathrm{Mul}\ (L^1(\mathbb{R}^+))$ is a strongly continuous one parameter semigroup from $\mathbb{R}^+$ into the multiplier algebra of $L^1(\mathbb{R}^+)$. Let $t \mapsto a^t : (0,\infty) \to L^1(\mathbb{R}^+)$ be a continuous semigroup in $L^1(\mathbb{R}^+)$ such that $(a^t * L^1(\mathbb{R}^+))^- = L^1(\mathbb{R}^+)$ and $\|a^t\|_1 \le 1$ for all $t > 0$; for example, let a^t be the fractional integral semigroup. Then $t \mapsto \delta^t * a^t : (0,\infty) \to L^1(\mathbb{R}^+)$ is a semigroup and is easily seen to be continuous, using the strong continuity and boundedness of δ^t, and the continuity and boundedness of a^t. Further $\delta^t * a^t * f \to f$ as $t \to 0$, $t > 0$. Also $t \mapsto \delta^t * a^t : (0,\infty) \to L^1(\mathbb{R}^+,\omega)$ is a continuous semigroup. Calculating the norm in $L^1(\mathbb{R}^+,\omega)$ we obtain

$$\begin{aligned}\|\delta^t * a^t\| &= \int_0^\infty |\delta^t * a^t(x)|\ \omega(x)\ dx \\ &= \int_t^\infty |a^t(x - t)|\ \omega(x)\ dx \\ &\le \omega(t) \int_t^\infty |a^t(x - t)|\ dx \\ &\le \omega(t) \le g(t) \quad \text{for all } t \ge 2.\end{aligned}$$

5.5 GROWTH ON VERTICAL LINES - A TAUBERIAN THEOREM

In this section we shall show that restricting the growth of $\|a^t\|$ on a vertical line restricts the structure of the algebra. Throughout this section let $t \mapsto a^t : H \to A$ be an analytic semigroup from the open right half plane into a Banach algebra A. To show how the restrictions arise we begin with an elementary result. Suppose that $|y|^{-1} \log \|a^{1+iy}\|$ tends to zero as $|y|$ tends to infinity $(y \in \mathbb{R})$. Let ϕ be a character on a maximal commutative subalgebra of A containing the semigroup a^t. Then there is a $\beta \in \mathbb{C}$ such that $\phi(a^t) = \exp(t\beta)$ for all $t \in H$, because $t \mapsto \phi(a^t) : H \to \mathbb{C}$ is a continuous semigroup. Now

$$\begin{aligned} & |y|^{-1} (\operatorname{Re} \beta - y \operatorname{Im} \beta) \\ &= |y|^{-1} \operatorname{Re} ((1 + iy)\beta) \\ &= |y|^{-1} \log |\phi(a^{1 + iy})| \\ &\le |y|^{-1} \log \|a^{1 + iy}\| \end{aligned}$$

tends to zero as $|y|$ tends to infinity. Thus $\operatorname{Im} \beta = 0$, and so $\sigma(a^t) \subseteq \mathbb{R}$ for all $t > 0$.

In Theorem 5.6 we shall see that a stronger restriction on the growth of $\|a^{1+iy}\|$ has fundamental implications for the structure of the Banach algebra. First we recall some definitions. If $\alpha > 0$, then $\log^+\alpha = \log \alpha$ for $\alpha \ge 1$, and $\log^+\alpha = 0$ for $0 < \alpha \le 1$. Then $\log^+$ is a continuous increasing function on $(0,\infty)$, and $\log^+ (\alpha\beta) \le \log^+\alpha + \log^+\beta$ for all $\alpha,\beta > 0$. An ideal J in a Banach algebra is called a <u>primitive ideal</u> if there is an <u>algebraically irreducible representation</u> θ from A into $BL(X)$ for a Banach space X such that $\ker \theta = J$. Note that in a Banach algebra each primitive ideal is closed. If A is a commutative Banach algebra, then an ideal J is primitive if and only if it is the kernel of character (that is, J is a maximal modular ideal), because the characters are the only irreducible representations of the algebra. Note that the probability Lebesque measure $\frac{1}{2\pi} d\theta$ on the unit circle $T = \{z \in \mathbb{C} : |z| =$ is mapped to the measure $\frac{1}{\pi} \frac{dy}{1+y^2}$ on $i\mathbb{R}$ by the conformal mapping $z \mapsto \frac{z + 1}{1 - z} : \mathbb{D} \to H$, where $\mathbb{D}$ is the open unit disc $\{z \in \mathbb{C} : |z| < 1\}$ in $\mathbb{C}$.

5.6 <u>THEOREM</u>

Let A be a Banach algebra, and let $t \mapsto a^t : H \to A$ be an analytic semigroup from the open right half plane H into A such that $(a^tA)^- = A$ for all $t \in H$. If $\int_{\mathbb{R}} (1 + y^2)^{-1} \log^+ \|a^{1+iy}\| \, dy$ is finite, then each proper closed (two sided) ideal in A is contained in a primitiv ideal.

We begin with a lemma that ensures that by moving to the right in the complex plane we obtain functions of exponential type.

5.7 LEMMA

If $t \mapsto a^t : H \to A$ is an analytic semigroup satisfying $\int_{\mathbb{R}} (1 + y^2)^{-1} \log^+ \|a^{1+iy}\| \, dy$ is finite, then $t \mapsto a^{3+t} : H \to A$ is of exponential type and $\int_{\mathbb{R}} (1 + t^2)^{-1} \log^+ \|a^{3+iy}\| \, dy$ is finite.

Proof. The second conclusion follows from the subadditivity of $\log^+$ and the finiteness of the integral $\int_{\mathbb{R}} (1 + y^2)^{-1} \, dy$. We consider the first conclusion which is proved by using the ideas behind the result that the difference of a set of positive measure contains an interval. Let $M = \int_{\mathbb{R}} (1 + y^2)^{-1} \log^+ \|a^{1+iy}\| \, dy$ and choose a positive real number m so that $e^{-M/m} = 3/4$. Let $V = \{y \in \mathbb{R} : \|a^{1+iy}\| \le e^{m|y|}\}$, and let μ be Lebesque measure on $\mathbb{R}$. Then V is a closed subset of $\mathbb{R}$, and $M \ge \int m|y| \, (1 + y^2)^{-1} \, dy$ where the integral is evaluated over the set $[-\beta,\beta] \setminus (V \cup [-1,1])$ for β a large positive real number. Note that $\log^+ \|a^{1+iy}\| \ge m|y|$ for all $y \notin V$. Now the function $y \mapsto y(1 + y^2)^{-1} : \mathbb{R}^+ \to \mathbb{R}^+$ is positive, increasing on $[0,1]$, and decreasing on $[1,\infty)$. Using the symmetry of the integral about 0 and considering the worst position of V with respect to $[-\beta,\beta]$, we have

$$M \ge 2\, m \int_{\alpha+1}^{\beta} y\,(1 + y^2)^{-1} \, dy$$

where $2\alpha = \mu(V \cap [-\beta,\beta])$. Hence

$$M \ge m \int_{\alpha+1}^{\beta} y^{-1} \, dy = m \log(\beta(\alpha + 1)^{-1})$$

so that $e^{M/m} \ge \beta(\alpha + 1)^{-1}$ and $\alpha \ge 3\beta/4 - 1$.

Let $w \in \mathbb{R}$ with $|w| \ge 5$, and let $\beta = 2|w|$, and suppose that

$$w \notin (V \cap [-\beta,\beta]) + (V \cap [-\beta,\beta]).$$

Then

$$(V \cap [-\beta,\beta] - w) \cap (V \cap [-\beta,\beta]) = \emptyset,$$

so that

$$\begin{aligned} 2\beta &\ge \mu((V \cap [-\beta,\beta] - w) \cap [-\beta,\beta]) + \mu(V \cap [-\beta,\beta]) \\ &\ge 2\alpha - |w| + 2\alpha \\ &\ge 3\beta - 4 - \beta/2. \end{aligned}$$

Therefore $4 \geq \beta/2$ contrary to the choice of w and β. Hence there are $y_1, y_2 \in V \cap [-\beta\ \beta]$ such that $w = y_1 + y_2$, so

$$\begin{aligned}\|a^{2+iw}\| &\leq \|a^{1+iy_1}\|.\|a^{1+iy_2}\| \\ &\leq e^{m|y_1|}.\ e^{m|y_2|} \\ &\leq e^{4m|w|}.\end{aligned}$$

The function $y \mapsto \|a^{1+iy}\| : [-5,5] \to \mathbb{R}$ is continuous, and so there is a constant C such that $\|a^{2+iw}\| \leq C\, e^{4m|w|}$ for all $w \in \mathbb{R}$.

Because $t \mapsto a^t : (0,\infty) \to A$ is a continuous semigroup there are constants m_1 and C_1 such that $\|a^t\| \leq C_1\, e^{m_1 t}$ for all $t \geq 1$. If $z = x + iy \in H$, then

$$\begin{aligned}\|a^{3+z}\| &\leq \|a^{1+x}\|.\ \|a^{2+iy}\| \\ &\leq C\ C_1\ e^{m_1(1+x)}.e^{4m|y|} \\ &\leq C\ C_1\ e^{m_1}.\exp\ (\max\ \{m_1, 4m\}.2^{1/2}\ |z|)\end{aligned}$$

because $x + |y| \leq 2^{1/2}\ |z|$ by Cauchy's inequality. Thus $z \mapsto a^{3+z} : H \to A$ is analytic of exponential type.

Note that in the above lemma, and in Theorem 5.6 we do not need an assumption on the growth of $\|a^t\|$ as $t \to 0$, $t > 0$. Perhaps if there is a semigroup satisfying the hypotheses of Theorem 5.6, then there is one that is bounded on $(0,1]$.

5.8 PROOF OF THEOREM 5.6

We begin by showing that A is not a radical algebra from which the result will follow by standard Banach algebra techniques. Using the Hahn-Banach Theorem we choose $F \in A^*$ such that $\|F\| = 1$ and $F(a^3) \neq 0$. We let $f(z) = F(a^{3+z})$ for all z in a neighbourhood of H^-. Then f is analytic in a neighbourhood $(H - 1$, say) of H^-, is of exponential type in H, and satisfies

$$\int_{\mathbb{R}} (1 + y^2)^{-1} \log^+ |F(iy)|\,dy \leq \int_{\mathbb{R}} (1 + y^2)^{-1} \log^+ \|a^{3+iy}\|\,dy$$

is finite by Lemma 5.7. By the Ahlfors-Heins Theorem (A1.1) there is a constant c such that $r^{-1} \log |F(re^{i\theta})| \to c \cos\theta$ as $r \to \infty$ for almost all $\theta \in (-\pi/2, \pi/2)$. However $r^{-1} \log |F(re^{i\theta})| \le r^{-1} \log \|a^3\| + r^{-1}$ $\log \|a^{re^{i\theta}}\|$ tends to minus infinity as r tends to infinity, because $r \mapsto a^{re^{i\theta}} : (0,\infty) \to A$ is a continuous homomorphism into a radical Banach algebra. This gives a contradiction, and so A is not a radical Banach algebra.

Finally let J be a proper closed ideal in A, and suppose that J is contained in no primitive ideal in A. Then A/J is a radical Banach algebra, and $t \mapsto a^t + J : H \to A/J$ is an analytic semigroup with

$$\int_{\mathbb{R}} (1 + y^2)^{-1} \log^+ \|a^{1+iy} + J\| \, dy$$

finite. This contradicts what we have just proved, and so A/J is not radical. Thus there is a primitive ideal in A containing J.

5.9 COROLLARY

If J is a proper closed ideal in $L^1(\mathbb{R}^n)$, then there is a $w \in R^n$ such that $f^\wedge(w) = 0$ for all $f \in J$.

Proof. The Gaussian semigroup (2.15) or Poisson semigroup (2.17) in $L^1(\mathbb{R}^n)$ satisfy the hypotheses of Theorem 5.6. A character on $L^1(\mathbb{R}^n)$ is of the form $g \mapsto g^\wedge(w)$ for some $w \in \mathbb{R}^n$ because of the identification of the carrier space of $L^1(\mathbb{R}^n)$ with $\mathbb{R}^n$ and of the Gelfand transform with the Fourier transform.

5.10 PROBLEM

Let G be a locally compact group. The group G is said to be of polynomial growth, or $G \in [PG]$, if and only if for each compact subset $W \subseteq G$ there is a positive integer r and a $C > 0$ such that $\mu(W^n) \le Cn^r$, where μ is left Haar measure on G and $W^n = \{w_1.\cdots.w_n : w_j \in W\}$. Does a metrizable locally compact group G have polynomial growth if and only if there is an analytic semigroup $t \mapsto b^t : H \to L^1(G)$ and a positive integer N such that

(i) $(b^t * L^1(G))^- = L^1(G) = (L^1(G) * b^t)^-$ for all $t \in H$,

(ii) $\|b^t\|_1 = 1$ and $b^t \ge 0$ as a function and in the *-algebra $L^1(G)$ for all $t > 0$, and

(iii) $\|b^{1+iy}\|_1 = O(|y|^N)$ as $|y| \to \infty$, $y \in \mathbb{R}$?

See Hulanicki [1974] and Dixmier [1960].

5.11 COROLLARY

Let $t \mapsto a^t : H \to A$ be an analytic semigroup into a Banach algebra A with $(a^t A)^- = A$ for all $t \in H$. If there is a non-zero continuous homomorphism from A into a radical Banach algebra, then $\int_{\mathbb{R}} (1 + y^2)^{-1} \log^+ \|a^{1 + iy}\| \, dy = \infty$.

Proof. Let $B = (\theta(A))^-$ where θ is a continuous homomorphism from A into a radical Banach algebra. Then $t \mapsto b^t = \theta(a^t) : H \to B$ is an analytic semigroup, and
$$\int_{\mathbb{R}} (1 + y^2)^{-1} \log^+ \|b^{1 + iy}\| \, dy$$
$$\leq \log^+ \|\theta\| \int_{\mathbb{R}} (1 + y^2)^{-1} \, dy + \int_{\mathbb{R}} (1 + y^2)^{-1} \log^+ \|a^{1 + iy}$$

Since B is a radical algebra the first integral diverges by Theorem 5.6, and hence so does the required integral.

5.12 SEMIGROUPS OF EXPONENTIAL TYPE - NONSEPARABILITY OF THE MULTIPLIER ALGEBRA

The basic result in this section is that if there is an analyti semigroup $t \mapsto a^t : H \to A$ into a commutative Banach algebra A such that $(a^t A)^- = A$ for all $t \in H$ and $t \mapsto a^t$ is of exponential type on H, then the multiplier algebra $\mathrm{Mul}(A)$ of A is nonseparable. The motivatinc example is $C_o(\mathbb{R})$, which has a bounded analytic semigroup defined in H and whose multiplier algebra, the algebra $C_b(\mathbb{R})$ of continuous bounded functions on $\mathbb{R}$, is non-separable. In this example the group of invertible elements in the multiplier algebra $C_b(\mathbb{R})$ has $2^{\aleph_0}$ connected components. To obtain these conclusions we require restrictions on the semigroup, and c the carrier space of $A^\#$ to get the $2^{\aleph_0}$ - components. Later in the sectic we show that some restrictions are necessary by means of examples. Before proving the main theorem of this section we shall state and prove a technical number theoretic lemma, and note some useful things about semi-groups and the multiplier algebra.

5.13 LEMMA

Let V be an infinite subset of the positive integers $\mathbb{N}$. The there is a set W of real numbers of cardinality $2^{\aleph_0}$ such that for each pair $\alpha \neq \beta \in W$ there is $n \in V$ with

$$|\exp(-2\pi i n\alpha) - \exp(-2\pi i n\beta)| \geq 1.$$

Proof. From V we choose an infinite subsequence m_j such that $m_{j+1} \geq 5m_j$ for all $j \in \mathbb{N}$. If T is a subset of $\mathbb{N}$, we let χ_T denote the characteristic function of T so that χ_T is 1 on T and 0 off T, and we let

$$\omega(T) = \frac{1}{2} \sum_{j=1}^{\infty} \frac{\chi_T(j)}{m_j}$$

Note that the series defining $\omega(T)$ converges because $m_j \geq 5^j$ for all j. Now we let $W = \{\omega(T) : T \subseteq \mathbb{N}\}$.

Let T and R be subsets of $\mathbb{N}$ with $T \neq R$, and let n be the least integer in the symmetric difference $(R\backslash T) \cup (T\backslash R)$. If

$$\gamma = \sum_{j>n} \frac{n(\chi_T(j) - \chi_R(j)),}{m_j}$$

then

$$\begin{aligned} |\gamma| &\leq \sum_{j>n} n \frac{2}{m_j} \\ &\leq \sum_{j>n} 2n.5^{-j} \\ &\leq n\, 2^{-1}5^{-n} \leq 2^{-1}. \end{aligned}$$

Since $\chi_T(j) = \chi_R(j)$ for $1 \leq j < n$, we have

$$\begin{aligned} &|\exp(-2\pi n i\, \omega(T)) - \exp(-2\pi n i\, \omega(R))| \\ &= |1 - \exp\{\pm \pi i + \pi i\gamma\}| \end{aligned}$$

with $+$ if $n \in T\backslash R$ and $-$ if $n \in R\backslash T$. Since $|\pi\gamma| \leq \pi/2$, $|1 + \exp \pi i\gamma| \geq 1$, and so $|\exp(-2\pi n i\, \omega(T)) - \exp(-2\pi n i\, \omega(R))| \geq 1$. This proves the lemma.

We note that the "≥ 1" in the lemma may be replaced by "$\geq 2 - \varepsilon$" if we replace the 5 in the proof by a large real number.

We return to semigroups and multiplier algebras. Note that if $t \mapsto a^t : H \to A$ is an analytic semigroup, then the semigroup is of exponential type (that is, there are constants C and K such that

$\|a^t\| \le C\exp(K|t|)$ for all $t \in H$) if and only if the set $\{\|a^t\| : t \in H, |t| \le 1\}$ is bounded. If A is a commutative Banach algebra with a bounded approximate identity bounded by 1, then the natural embedding of A into the multiplier algebra $\mathrm{Mul}(A)$ defined by $a \mapsto L_a : A \to \mathrm{Mul}(A)$, where $L_a(x) = ax$ for all $x \in A$, is an isometric monomorphism. We shall regard A as a closed ideal in $\mathrm{Mul}(A)$ via this embedding. A semigroup $t \mapsto b^t$ from an additive subsemigroup of $\mathbb{C}$ into $\mathrm{Mul}(A)$ is said to be strongly continuous if the map $t \mapsto b^t x$ is continuous into A for all $x \in A$. (See Chapter 6 for more on strongly continuous semigroups.)

5.14 THEOREM

Let A be a commutative Banach algebra without identity, and let $t \mapsto a^t : H \to A$ be an analytic semigroup with $(a^1A)^- = A$. If $\{\|a^t\| : t \in H, |t| \le 1\}$ is bounded, then there is a one parameter group $y \mapsto a^{iy} : \mathbb{R} \to \mathrm{Mul}(A)$ such that a^0 is the identity of $\mathrm{Mul}(A)$, and $t \mapsto a^t : H^- \to \mathrm{Mul}(A)$ is a strongly continuous norm discontinuous semigroup. Further $\mathrm{Mul}(A)$ is non-separable. If the carrier space of $A^\#$ has only finitely many components, then in the norm topology the group of invertible elements in $\mathrm{Mul}(A)$ has at least $2^{\aleph_0}$ components.

Proof. We first extend the semigroup $t \mapsto a^t : H \to \mathrm{Mul}(A)$ to a semigroup $t \mapsto a^t : H^- \to \mathrm{Mul}(A)$ by strong continuity. The boundedness hypothesis implies that the semigroup is of exponential type, and so is bounded on each bounded subset of H. Let $y \in \mathbb{R}$, $h \in A$, and $\varepsilon > 0$. For all $k \in A$ and $s, t \in H$, we have

$$\|a^t.h - a^s.h\| \le (\|a^t\| + \|a^s\|)\,\|h - a^1.k\| + \|a^{1+t}.k - a^{1+s}.k\|.$$

Since $(a^1A)^- = A$, we may choose k so that $\|h - a^1.k\|$ is very small, and then for s and $t \in H$ with $|iy - s|$ and $|iy - t|$ very small $\|a^{1+t}.k - a^{1+s}.k\|$ is very small. Hence $a^t.h$ converges in A as t converges in H to iy for each $y \in \mathbb{R}$ and each $h \in A$. We define the operator a^{iy} on A by $a^{iy}(h) = \lim a^t.h$ as t tends to iy, $t \in H$. From this definition and the properties of the semigroup $t \mapsto a^t : H \to A$, direct calculations yield that $t \mapsto a^t : H \to \mathrm{Mul}(A)$ is a strongly continuous semigroup and is bounded on bounded subsets of H^-. Also a^0 is the identity operator on A because $a^0(a^1k) = a^1k$ and $(a^1A)^- = A$.

If $t \mapsto a^t : H^- \quad Mul(A)$ were continuous, then there would be a small $t > 0$ such that $\|a^t - a^0\| = \|a^t - 1\| < 1$. Thus a^t would be invertible in $Mul(A)$, and so A would contain the identity of $Mul(A)$ contrary to hypothesis.

Let G denote the group of invertible elements in $Mul(A)$, and let r and s be distinct real numbers. Suppose that a^{ir} and a^{is} are in the same component of G. Then $y = r - s \neq 0$ and a^{iy} is the principal component of G. Since $Mul(A)$ is commutative (as we may check using $A^2 = A$) there is an element $c \in Mul(A)$ such that $a^{iy} = \exp c$. Let $t \mapsto b^t : H^- \to Mul(A)$ be defined by $b^t = a^{yt/2\pi}.\exp(itc/2\pi)$. Then $b^{2\pi i} = 1$ and $b^t \in A$ for all $t \in H$. If ϕ is a character on A, then $t \mapsto \phi(b^t) : H \to \mathbb{C}$ is a continuous semigroup and is non-zero because $(b^1A)^- = (a^1A)^- = A$. Hence there is a complex number β such that $\phi(b^t) = \exp(\beta t)$ for all $t \in H$. Now $\exp \beta(2\pi i + 1) = \phi(b^{2\pi i + 1}) = \phi(b^1)$ $= \exp \beta$ so that β is an integer. Also $\exp \beta = |\phi(b^1)| \leq \|b^1\|$ and $\beta \leq \log \|b^1\|$. Let U be the set of integers n such that there is a character ϕ on A with $\phi(b^t) = \exp(-nt)$ for all $t \in H$. Then U is non-empty and is bounded below by $-\log \|b^1\|$. Because A does not have an identity the spectrum $\sigma(b^1)$ of the element b^1 in A is $\{0\} \cup \{\exp(-nt) : n \in U\}$ for all $t \in H$.

Assume that $\sigma(b^1)$ is finite. Then $0 \in \sigma(b^1)$ and $\sigma(b^1) \neq \{0\}$. By the single variable analytic functional calculus for b^1 there is an idempotent e in the unital Banach subalgebra B of A generated by b^1 such that if ϕ is a character on B, then $\phi(e) = 1$ if $\phi(b^1) \neq 0$ and $\phi(e) = 0$ if $\phi(b^1) = 0$. Note that since the spectrum of b^1 is finite the unital Banach algebra generated by b^1 is the same as that generated by b^1 and $(b^1 - \lambda 1)^{-1}$ for all λ in $\mathbb{C} \setminus \sigma(b^1)$. If ϕ is a character on A, then $\phi(b^1) \neq 0$ so $\phi(e) = 1$. Hence $(1 - e)A$ is a radical Banach algebra, and $t \mapsto (1 - e)b^t : H \to (1 - e)A$ is an analytic semigroup. Since $t \mapsto a^t$ and $t \mapsto \exp(tc)$ are bounded on bounded subsets of H^-, there is a k such that $\|b^{iy}\| \leq k$ for all $y \in [0, 2\pi]$. Using the 2π - periodicity of $y \mapsto b^{iy}$, it follows that $\{\|(1 - e)b^{1+iy}\| : y \in \mathbb{R}\}$ is bounded. By Theorem 5.6 it follows that $(1 - e)b^1 = 0$, and so $e = 1$, which is a contradiction.

Either $\sigma(b^1)$ is infinite or our supposition that a^{ir} and a^{is} are in the same component of G is wrong. If the carrier space of $A^{\#}$ has at most a finite number of components, then $\sigma(b^1)$ is finite and so G has at least $2^{\aleph_0}$ components. We now show that in this case $Mul(A)$ is nonseparable. There is a constant α such that $\|a^t\| \le \alpha$ for all $t \in i\mathbb{R}$ with $|t| \le 1$. Since $\|1 - a^{i(r-s)}\| \ge 1$, we have $\|a^{ir} - a^{is}\| \ge \alpha^{-1}$ for all $r, s \in \mathbb{R}$ with $|r| \le 1$, $|s| \le 1$. Thus a dense subset of Mul(A) has at least the cardinality of $[-1,1]$.

To complete the proof we show that Mul(A) is nonseparable if $\sigma(b^1)$ is infinite. In this case $U = \{n \in \mathbb{N} : \exp(-nt) \in \sigma(b^t)$ for all $t \in H\}$ is infinite. Because U is bounded below by $-\log\|b^1\|$, the set $V = U \cap \mathbb{N}$ is infinite. Let W be the set of cardinality $2^{\aleph_0}$ correspondi[ng] to V given by Lemma 5.13. Let $\alpha, \beta \in W$ with $\alpha \ne \beta$. Then there is $n \in V$ such that

$$|\exp(-2\pi n i\alpha) - \exp(-2\pi n i\beta)| \ge 1,$$

and there is a character ϕ on $A^{\#}$ such that $\phi(b^t) = \exp(-nt)$ for all $t \in H$. We now extend the character ϕ from A to Mul(A) by a standard process. For each $d \in$ Mul(A) the linear functional $x \mapsto \phi(dx) : A \to \mathbb{C}$ has kernel containing the kernel of ϕ because $\phi(x) = 0$ and $\phi(z) \ne 0$ with $z \in A$ implies that $\phi(dx)\,\phi(z) = \phi(dxz) = \phi(dz)\,\phi(x) = 0$. Hence the functional $x \mapsto \phi(dx)$ is a constant multiple of ϕ, and we denote this constant by $\tilde\phi(d)$ so that $\tilde\phi(d)\,\phi(x) = \phi(dx)$ for all $d \in$ Mul(A) and $x \in A$. From this equation it follows that $\tilde\phi$ is a character on Mul(A), and that $\tilde\phi$ restricted to A is equal to ϕ. Therefore $\tilde\phi(b^{iy})\,\tilde\phi(b^1) = \phi(b^{1+iy}) = \exp(-n(1+iy)) = \exp(-niy)\,\phi(b^1)$, and $\tilde\phi(b^{iy}) = \exp(-niy)$ for all $y \in \mathbb{R}$. Hence

$$\begin{aligned} &\|b^{2\pi i\alpha} - b^{2\pi i\beta}\| \\ \ge\ &|\tilde\phi(e^{2\pi i\alpha}) - \tilde\phi(e^{2\pi i\beta})\| \\ =\ &|\exp(-2\pi n i\alpha) - \exp(-2\pi n i\beta)| \\ \ge\ &1. \end{aligned}$$

So Mul(A) is nonseparable and the proof is complete.

5.15 EXAMPLE

Note that we cannot deduce that G has uncountably many components without the hypothesis that the carrier space of $A^{\#}$ has only a finite number of components. Here is an example to show this. Take $A = c_o$ so that Mul(A) $= \ell^\infty$. Then there is a semigroup in A satisfying the hypotheses of the theorem. For example, let $a^t = (n^{-t})$ for all $t \in H$.

However the group of invertible elements in ℓ^∞ is connected because if $(\alpha_n) \in G$, then $(\alpha_n) = \exp((\log|\alpha_n| + i \operatorname{Arg} \alpha_n))$ where Arg is the principal value of the argument. We are using the observation that if (α_n) is invertible, then $\{|\alpha_n| : n \in \mathbb{N}\}$ is bounded away from zero.

5.16 COROLLARY

Let A be a commutative Banach algebra without identity and with only a finite number of components in the carrier space of A. If A is a quotient of a uniform algebra with a countable bounded approximate identity, then the group of invertible elements in $\mathrm{Mul}(A)$ has $2^{\aleph_0}$ components.

Proof. Let B be the uniform algebra of which A is the quotient. By Theorem 3.1 there is an analytic semigroup $t \mapsto b^t : H \to B$ such that $(b^t B)^- = B$ and $\nu(b^t) \leq \exp(\pi|\mathrm{Im}\, t|/2)$ for all $t \in H$, where ν is the spectral radius. Since the norm in B is equal to the spectral radius, the hypotheses of Theorem 5.14 are satisfied by the semigroup $t \mapsto b^t + J : H \to A = B/J$. This completes the proof.

5.17 EXAMPLE

We shall now give an example of a Banach algebra satisfying the hypotheses of the above corollary. Let $\mathfrak{A}$ be the algebra of functions continuous in the closed right half plane H^-, analytic in the open half plane H, and tending to zero at infinity in the sense that $|f(z)| \to 0$ as $|z| \to \infty$, $z \in H^-$. With the uniform norm on H^-, $\mathfrak{A}$ is a separable uniform algebra isometrically isomorphic to the maximal ideal $\{g \in A(\mathbb{D}) : g(1) = 0\}$ in the disc algebra $A(\mathbb{D})$. Using the maximal modulus theorem on large semidiscs $\{z \in H^- : |z| \leq n\}$, we see that $\|f\| = \sup\{|f(iy)| : y \in \mathbb{R}\}$ for all $f \in \mathfrak{A}$. For all $n \in \mathbb{N}$ and all $z \in H^-$, let $e_n(z) = n(n+z)^{-1}$, then $e_n \in \mathfrak{A}$ and $\|e_n\| \leq 1$. Further e_n is easily seen to be an approximate identity in $\mathfrak{A}$, because $e_n(iy)$ tends to 1 uniformly in compact subsets of $\mathbb{R}$ as n tends to infinity. A quotient A of $\mathfrak{A}$ with at most a finite number of components in the carrier space of $A^\#$ will satisfy the hypotheses of Corollary 5.16. Let $g(z) = e^{-z}$ for all $z \in H^-$. Then multiplication by g is a multiplier on $\mathfrak{A}$ and $\|gf\|_\infty = \|f\|_\infty$ for all $f \in \mathfrak{A}$. Hence $g\mathfrak{A}$ is a proper closed ideal in $\mathfrak{A}$. A character ϕ on $\mathfrak{A}$ has the form $\phi(f) = f(t)$ for some $t \in H^-$ and all $f \in \mathfrak{A}$. Since $g(t) \neq 0$ for all $t \in H^-$, no character on $\mathfrak{A}$ annihilates $g\mathfrak{A}$. Thus $A = \mathfrak{A}/g\mathfrak{A}$ is a radical Banach algebra and satisfies the hypotheses of Corollary 3.16 because the carrier space of $A^{\#}$ is a single point.

5.18 EXAMPLE

We shall construct an example of a separable Banach algebra A with a countable bounded approximate identity with multiplier algebra $Mul(A)$ equal to $A \oplus \mathbb{C}1$. This shows that some hypothesis on the semigrou is essential for the conclusions of Theorem 5.14. Let bv be the set of complex sequences (α_n) such that

$$\|(\alpha_n)\| = \sup |\alpha_n| + \sum_1^{\infty} |\alpha_{n+1} - \alpha_n|$$

is finite. With this norm and coordinatewise algebraic operations, bv is a Banach space. An elementary calculation using the inequality

$$\sum_1^{m} |\alpha_{n+1}\beta_{n+1} - \alpha_n\beta_n|$$

$$\leq \sum_1^{n} (|\beta_{n+1}|.|\alpha_{n+1} - \alpha_n| + |\alpha_n|.|\beta_{n+1} - \beta_n|)$$

shows that bv is a unital Banach algebra with the coordinatewise product Let bv_o denote the set of (α_n) in bv such that α_n tends to zero as n tends to infinity. Since lim is a character on bv, bv_o is a closed maximal ideal in bv, and $bv = bv_o \oplus \mathbb{C}1$ though the usual norm on $bv_o \oplus \mathbb{C}1$ is only equivalent to that on bv. If we replace the sup in the definition of the norm on bv by lim sup, then the norms match here but partial summation is needed to show bv is a Banach algebra.

We shall show that, if $A = bv_o$, then $Mul(A)$ is isomorphic (bicontinuously) to bv, and that A has a countable bounded approximate identity. Let f_n denote the sequence with a 1 in the n-th place and zeros elsewhere, and let $e_n = \sum_1^n f_j$. Then $f_n, e_n \in bv_o$ and $\|e_n\| = 2$ for all n. A direct calculation shows that $\{e_n : n \in \mathbb{N}\}$ is a bounded approximate identity in A. For each $(\alpha_n) \in bv$ define $L_{(\alpha_n)}$ on bv_o by $L_{(\alpha_n)}(\beta_n) = (\alpha_n\beta_n)$. Then $(\alpha_n) \mapsto L_{(\alpha_n)}$ is a bicontinuous monomorphism from bv into $Mul(bv_o)$. If $T \in Mul(bv_o)$, then $T(f_m) = T(f_m^{\ 2}) = T(f_m)$ so that there is a complex sequence (γ_n) such that $T(f_m) = \gamma_m f_m$ for al

m. Hence

$$\max\{|\gamma_n| : 1 \le n \le m\} + \sum_{1}^{m-1} |\gamma_{n+1}-\gamma_n| + |\gamma_m| = \|T(e_m)\| \le 2\|T\|$$

for all $m \in \mathbb{N}$ so that $(\gamma_n) \in bv$. Further $Te_m = L_{(\gamma_n)} e_m$ for all $m \in \mathbb{N}$. Since T and $L_{(\gamma_n)}$ are continuous multipliers, and $\{e_m : m \in \mathbb{N}\}$ is a bounded approximate identity, we have $T = L_{(\gamma_n)}$. This completes the properties of the example.

The algebra bv_o is not a radical algebra. Does there exist a separable commutative radical Banach algebra A with a bounded approximate identity such that $Mul(A) = A \oplus \mathbb{C}\,1$?

5.19 NOTES AND REMARKS ON CHAPTER 5

Remarks on 5.2 - growth on rays

The results in this section are from Esterle [1980e]. Theorems 5.3 and 5.4 are Theorems 3.1 and 3.6 of Esterle [1980e]. However our version of Theorem 5.4 is weaker than Esterle's in that his semigroup is infinitely differentiable. The infinite differentiability of the semigroup $t \mapsto \delta^t * a^t$ comes by choosing an infinitely differentiable semigroup a^t in $L^1(\mathbb{R}^+)$ such that a^t and all its derivatives are in $\bigcap_1^\infty \mathrm{Dom}\,(D^n)$, where D (= differentiation) is the infinitesimal generator of the strongly continuous semigroup $t \mapsto \delta^t$ on $L^1(\mathbb{R}^+)$. For example the semigroup $t \mapsto c^t$ of 2.9 has the required properties. See Esterle [1980e] for details. We shall now mention related results. If $t \mapsto a^t : [0,\infty) \to A$ is a semigroup that has an extension to an analytic semigroup in an open neighbourhood of $(0,\infty)$ in $\mathbb{C}$, then there exists a $\lambda > 0$ such that $\lim_{r\to\infty} \exp(-\lambda r).\log\|a^r\| = 0$ (see Esterle [1980e] Theorem 3.3).

Esterle [1980e] also investigates the rates of decrease of $\|a^n\|^{1/n}$ for a in a radical Banach algebra using various general methods. Bade and Dales [1981] study similar problems for the radical algebras $L^1(\mathbb{R}^+,\omega)$ providing specific rates of growth depending on ω. There are further results on the rates of growth of $\|a^n\|^{1/n}$ in Esterle [1980d].

Remarks on 5.5 - growth on vertical lines

The results in this section are in Esterle [1980f] though sometimes the minor details are a little different. Lemma 5.7 was suggested to me by A.M.Davie as a way of eliminating the hypothesis of exponential type. The following references are related to problem 5.10: Leptin [1973], [1976], Hulanicki [1974], and Dixmier [1960]. See also Dales and Hayman [1981].

Remarks on 5.12 - semigroups of exponential type

The results in this section are all from Esterle [1980c], and we have not discussed all the theorems in that paper. The discontinuity of the one parameter group $y \mapsto a^{iy} : \mathbb{R} \to \mathrm{Mul}(A)$ constructed in the proof of Theorem 5.14 leads quickly to the fact that $\mathrm{Mul}(A)$ is non-separable by the following result of Esterle's [1980c, Theorem 3.1].

THEOREM. Let X be a Banach space, and let $t \mapsto b^t : (0,\infty) \to BL(X)$ be a strongly continuous semigroup. If the set $\{b^t : t > 0\}$ is separable in the norm topology on $BL(X)$, then the semigroup is continuous in the norm topology.

The proof of this uses the separation of Borel sets by analytic sets (see Hoffman-Jørgensen [1970] Theorem 5, Section 2, Chapter 3) to show that the semigroup $t \mapsto b^t$ is measurable. From this the result follows by a standard theorem in the theory of one parameter semigroups (see Hille and Phillips [1974]). If the semigroup in the above theorem is actually a one parameter group, then the continuity of the semigroup may be proved by versions of the closed graph theorem for metric groups thereby avoiding the separation theorem and the result from semigroup theory. Example 5.18 is due to S.Grabiner [1980].

6 NILPOTENT SEMIGROUPS AND PROPER CLOSED IDEALS

6.1 INTRODUCTION

We know that an analytic semigroup $t \mapsto a^t : H \to A$ into a Banach algebra A has the property that $(a^t A)^- = (a^1 A)^-$ for all $t \in H$. In this Chapter we shall be concerned with continuous semigroups $t \mapsto a^t : (0,\infty) \to A$ satisfying $(\bigcup_{t>0} a^t A)^- = A$ and $(a^r A)^- \neq A$ for each $r > 0$. Clearly these semigroups are not analytic. However analyticity will play an important role later in this Chapter. In the first section the standard Hille-Yoshida Theorem is proved for strongly continuous contraction semigroups on a Banach space. There are excellent accounts of this theorem and some of its applications in Dunford and Schwartz [1958], Reed and Simon [1972], and Hille and Phillips [1974]. We have included it for completeness. From Corollary 6.9 on the results are less standard and involve Banach algebra conditions or the nilpotency of the semigroups. In the process we prove a hyperinvariant subspace theorem for a suitable quasinilpotent operator on a Banach space, and investigate when there is a norm reducing monomorphism from $L^1_*[0,1]$ into a Banach algebra. These results are due to J.Esterle and were given in detail in his 1979 U.C.L.A. lectures.

6.2 STRONGLY CONTINUOUS CONTRACTION SEMIGROUPS ON BANACH SPACES

In this section we introduce the notation and definitions required later in this chapter, and give a proof of the Hille-Yoshida Theorem. The simplest example underlying this theorem is that, if $t \mapsto a^t : (0,\infty) \to \mathbb{C}$ is a continuous semigroup with $|a^t| \leq 1$ for all $t > 0$, then there is an $R \in \mathbb{C}$ such that $\mathrm{Re}\, R \leq 0$ and $a^t = \exp(tR)$ for all $t > 0$. Here R is to be thought of as a linear operator on $\mathbb{C}$, and is the derivative of the semigroup at $t = 0$. This little result will be generalised by replacing $\mathbb{C}$ by $BL(X)$ for X a Banach space, and finding necessary and sufficient conditions on a closed linear operator R on the Banach space X for it to generate the semigroup in a suitable way. Throughout this section let X denote a Banach space.

6.3 DEFINITIONS

A closed (linear) operator R on a Banach space X is a line[ar] operator R defined on a dense linear subspace $\mathcal{D}(R)$ of X such that R has closed graph $\{(x,Rx) : x \in \mathcal{D}(R)\}$ in $X \times X$. The closed graph proper[ty] is equivalent to the condition that

$$x_n \in \mathcal{D}(R),\ x_n \to x \in X, \text{ and } Rx_n \to y \in X \text{ imply that}$$
$$x \in \mathcal{D}(R) \text{ and } Rx = y.$$

Note that each continuous linear operator on X is a closed linear operat[or] and that, using Zorn's lemma, one can easily construct linear operators on dense subspaces of an infinite dimensional Banach space that are not closed.

The following method of constructing closed operators lies beh[ind] the definition of the resolvent of a closed operator. Let T be a continu[ous] one-to-one linear operator on a Banach space X with TX dense in X. Define $R : Tx \mapsto x : TX \to X$. Then R is a closed linear operator with domain $\mathcal{D}(R) = TX$, because $x_n \in TX$, $x_n \to x \in X$, $R x_n \to y \in X$ imply that $x_n = TRx_n \to x = Ty$ so that $x \in TX$ and $Rx = T^{-1}x = y$. By direct calculation we could check that the Laplacian on $L^2(\mathbb{R}^n)$ and the derivative on $L^1(\mathbb{R})$ are closed operators.

The resolvent set $\rho(R)$ of a closed operator R on a Banach space X is the set of complex numbers λ such that $\lambda - R$ is a one-to-o[ne] linear operator from $\mathcal{D}(R)$ onto X whose algebraic inverse is continuous. The complement of $\rho(R)$ in $\mathbb{C}$ is the spectrum of R, and the function $\lambda \mapsto (\lambda - R)^{-1} : \rho(R) \to BL(X)$ is called the resolvent of R.

6.4 LEMMA

Let R be a closed operator on a Banach space X with non-empty resolvent set $\rho(R)$. Then the resolvent set $\rho(R)$ is open, and the map $\lambda \mapsto (\lambda - R)^{-1} : \rho(R) \to BL(X)$ is an analytic function into a commutative subset of $BL(X)$.

Proof. An elementary algebraic calculation shows that, if $\lambda, \mu \in \rho(R)$, th[en]

$$(\lambda - R)^{-1} - (\mu - R)^{-1} = (\mu - \lambda)(\lambda - R)^{-1} (\mu - R)^{-1}$$
$$= (\mu - \lambda)(\mu - R)^{-1} (\lambda - R)^{-1}.$$

The openness of the resolvent set of R follows from the corresponding result for bounded linear operators. Let $\lambda \in \rho(R)$ and let $\alpha \in \mathbb{C}$ with $|\alpha - \lambda| < \|(\lambda - R)^{-1}\|^{-1}$. Then

$$\alpha - R = (\lambda - R) + \alpha - \lambda$$
$$= (1 + (\alpha - \lambda)(\lambda - R)^{-1})(\lambda - R)$$

on $\mathcal{D}(R)$ so that $(\alpha - R)^{-1} \in BL(X)$ because $(\lambda - R)^{-1} \in BL(X)$ and $(1 + (\alpha - \lambda)(\lambda - R)^{-1})$ is an invertible bounded linear operator since $\|(\alpha - \lambda)(\lambda - R)^{-1}\| < 1$. Using the geometric series expansion of $(1 + (\alpha - \lambda)(\lambda - R)^{-1})^{-1}$, we obtain $\|(\alpha - R)^{-1}\| \le \|(\lambda - R)^{-1}\| (1 - |a - \lambda|.\| \lambda - R)^{-1}\|)^{-1}$. From the equation $(\lambda - R)^{-1} - (\alpha - R)^{-1} = (\alpha - \lambda)(\lambda - R)^{-1}(\alpha - R)^{-1}$ we deduce firstly that $\lambda \mapsto (\lambda - R)^{-1} : \rho(R) \to BL(X)$ is continuous, and then after dividing by $(\lambda - \alpha)$ that this function is analytic. This proves the lemma.

6.5 DEFINITIONS

A semigroup $t \mapsto b^t$ from $(0,\infty)$ into a Banach algebra is called a contraction semigroup if $\|b^t\| \le 1$ for all $t > 0$.

Most of the semigroups we constructed in Chapters 2 and 3 are contraction semigroups if attention is restricted to the positive real numbers. A function F from a topological space U into $BL(X)$ is strongly continuous if $t \mapsto F(t).x : U \to X$ is continuous for all $x \in X$. Clearly a (norm) continuous function from U into $BL(X)$ is strongly continuous.

Here are a couple of examples of semigroups that satisfy the hypotheses of the following lemma, which is part of the Hille-Yoshida Theorem.

If we take $X = L^2(\mathbb{R})$ and $(b^t f)(w) = f(w + t)$ for all $f \in L^2(\mathbb{R})$, $t \in \mathbb{R}$, and $w \in \mathbb{R}$, then $t \mapsto b^t : \mathbb{R} \to BL(L^2(\mathbb{R}))$ is a strongly continuous contraction (semi)group with $b^0 = I$. The closed operator $R = D$, where D is the differentiation operator, and the domain of R is the space of $f \in L^2(\mathbb{R})$ such that f is differentiable almost everywhere on $\mathbb{R}$ and $Df \in L^2(\mathbb{R})$.

If A is a Banach algebra with a continuous contraction semigroup $t \mapsto a^t : (0,\infty) \to A$ satisfying $(\bigcup_{t>0} a^t A)^- = A$, then we take

$X = A$, b^0 = the identity operator on A, and $b^t(x) = a^t.x$ for all $t > 0$ and all $x \in A$. Then $t \mapsto b^t : [0,\infty) \to BL(A)$ is a strongly continuous contraction semigroup with $b^0 = I$. The generator R of the semigroup is a closed operator on A satisfying the multiplier equation $R(x.a) = R(x).a$ for all $x \in \mathcal{D}(R)$ and $a \in A$.

6.6 LEMMA

Let $t \mapsto b^t : [0,\infty) \to BL(X)$ be a strongly continuous contraction semigroup with $b^0 = I$. Let

$$\mathcal{D}(R) = \{x \in X : \lim_{\substack{t \to 0,\\ t>0}} t^{-1}(b^t - 1).x \text{ exists in } X\}.$$

Then $\mathcal{D}(R)$ is a dense linear subspace of X, and if $Rx = \lim t^{-1}(b^t - 1)$ for all $x \in \mathcal{D}(R)$, then R is a closed operator on X. Further the resolvent set $\rho(R)$ of R contains the open right half plane H, and $\|(\lambda - R)^{-1}\| \le (\mathrm{Re}\,\lambda)^{-1}$ for all $\lambda \in H$.

Proof. Clearly $\mathcal{D}(R)$ is a linear subspace of X. To show that it is dense we average $b^w.x$ over small intervals of the form $[0,s]$. If $x \in X$ and $s > 0$, then

$$(1)\quad t^{-1}(b^t - 1)\int_0^s b^w x\,dw = \int_0^s b^w\, t^{-1}(b^t - 1)\,x\,dw$$
$$= t^{-1}\int_0^s (b^{t+w} - b^w)x\,dw$$
$$= t^{-1}\int_s^{s+t} b^w\,x\,dw - t^{-1}\int_0^t b^r\,x\,dr$$
$$\to b^s x - x$$

as $t \to 0$, $t > 0$, because $w \mapsto b^w x : [0,\infty) \to X$ is a continuous function with $b^0 x = x$. Thus $\int_0^s b^w\,x\,dw \in \mathcal{D}(R)$ and $R\left(\int_0^s b^w\,x\,dw\right) = (b^s - 1)x$ for all $x \in X$ and all $s > 0$. Since $s^{-1}\int_0^s b^w\,x\,dw \to x$ as $s \to 0$, $s > 0$, it follows that $\mathcal{D}(R)$ is dense in X. If $x \in \mathcal{D}(R)$, then we also obtain from (1) that $\int_0^s b^w\,R\,x\,dw = (b^s - 1)\,x$ for all $s > 0$.

We shall use this equality to show that R has a closed graph. Let x_n be a sequence in $\mathcal{D}(R)$ such that $x_n \to x \in X$ and $Rx_n \to y \in X$

as $n \to \infty$. Then

$$\begin{aligned}
&s^{-1}(b^s - 1)x \\
&= \lim s^{-1}(b^s - 1)x_n \\
&= \lim s^{-1}\int_0^s b^w R x_n \, dw \\
&= s^{-1}\int_0^s b^w y \, dw \\
&\to y \quad \text{as} \quad s \to 0,\ s > 0
\end{aligned}$$

because the function $w \mapsto b^w y : [0,\infty) \to X$ is continuous with $b^0 y = y$. Hence $x \in \mathcal{D}(R)$ and $Rx = y$.

The Laplace transform of the function $w \mapsto e^{\alpha w} : [0,\infty) \to \mathbb{C}$ for $-\alpha \in H$ is the function $\lambda \mapsto (\lambda - \alpha)^{-1} : H^- \to \mathbb{C}$. This is the motivation behind the definition of $R(\lambda) = (\lambda - R)^{-1}$ given below. For each $\lambda \in H$ the operator $R(\lambda)$ on X is defined by

$$R(\lambda)x = \int_0^\infty e^{-\lambda w} b^w x \, dw$$

for each $x \in X$. Clearly the integral is defined and convergent for all $x \in X$, and $R(\lambda)$ is a linear operator on X. Further

$$\begin{aligned}
\|R(\lambda)x\| &\le \int_0^\infty e^{-w(\operatorname{Re}\lambda)} \|b^w\| \, \|x\| \, dw \\
&\le (\operatorname{Re}\lambda)^{-1} \|x\|
\end{aligned}$$

for all $x \in X$, so that $R(\lambda) \in BL(X)$. If $x \in X$ and $\lambda \in H$, then

$$\begin{aligned}
&t^{-1}(b^t - 1)\, R(\lambda)x \\
&= t^{-1}\int_0^\infty e^{-\lambda w}(b^{t+w} - b^w)\, x \, dw \\
&= t^{-1} e^{\lambda t}\int_t^\infty e^{-\lambda v} b^v x \, dv - t^{-1}\int_0^\infty e^{-\lambda w} b^w x \, dw \\
&= t^{-1}(e^{\lambda t} - 1)\int_0^\infty e^{-\lambda v} b^v x \, dv - e^{\lambda t} t^{-1}\int_0^t e^{-\lambda w} b^w x \, dw \\
&\to \lambda R(\lambda)x - x
\end{aligned}$$

as $t \to 0$, $t > 0$. Thus $R(\lambda)x \in \mathcal{D}(R)$, and $R\,R(\lambda)x = \lambda R(\lambda)x - x$ so that $(\lambda - R)\,R(\lambda)$ is the identity operator on X. If $x \in \mathcal{D}(R)$ and $\lambda \in H$, then

$$R(\lambda)\ t^{-1}\ (b^t - 1)x = t^{-1} \int_0^\infty e^{-\lambda w}\ (b^{t+w} - b^w)x\ dw$$

converges to $\lambda R(\lambda)x - x$ as $t \to 0,\ t > 0,$ as in the calculation above, and converges to $R(\lambda)Rx$ by the definition of R. Hence $R(\lambda)(\lambda - R)$ is the identity operator on $\mathcal{D}(R)$. Thus $R(\lambda)$ is equal to $(\lambda - R)^{-1}$ as a linear operator, and $(\lambda - R)^{-1}$ is in $BL(X)$, which completes the proof.

6.7 THEOREM (HILLE-YOSHIDA THEOREM)

Let R be a closed linear operator on a Banach space X. There is a strongly continuous contraction semigroup $t \mapsto b^t : (0,\infty) \to BL(X)$ with $b^0 = I$ satisfying $Rx = \lim_{t\to 0} t^{-1} (b^t - 1)x$ for all $x \in \mathcal{D}(R)$ if and only if $(\lambda - R)^{-1} \in BL(X)$ and $\|(\lambda - R)^{-1}\| \le \lambda^{-1}$ for all $\lambda > 0$. If R and $t \mapsto b^t$ satisfy the above conditions, then the open right half plane H is contained in the resolvent set $\rho(R)$ of R, $(\lambda - R)^{-1}x = \int_0^\infty e^{-\lambda w}\ b^w x\ dw$ for all $x \in X$, and $\lambda \in H$, $\|(\lambda - R)^{-1}\| \le (\mathrm{Re}\ \lambda)^{-1}$ for all $\lambda \in H$, and the function $\lambda \mapsto (\lambda - R)^{-1} : H \to BL(X)$ is analytic. Further for each $x \in X$

$$\|b^t x - \exp t(\lambda^2(\lambda - R)^{-1} - \lambda)x\|$$

tends to zero uniformly for t in compact subsets of $[0,\infty)$ as λ tends to infinity.

The operator R occurring in the Hille-Yoshida Theorem is called the generator or infinitesimal generator of the semigroup $t \mapsto b^t$. Lemma 6.6 gives half the above result. The heuristic motivation for the construction of b^t from R is that we want $b^t = \exp(t\,R)$ in a suitable sense. Formally $\lambda^2(\lambda - R)^{-1} - \lambda = R(1 - R/\lambda)^{-1}$ converges to R as λ tends to infinity and each $\lambda^2(\lambda - R)^{-1} - \lambda$ is a continuous linear operator. This is why we expect b^t to be a suitable limit of $\exp t(\lambda^2(\lambda - R)^{-1} - \lambda)$ as λ tends to infinity, where the exponential of a bounded linear operator is defined by the power series for the exponential.

Proof of the Hille-Yoshida Theorem.

Suppose that $(\lambda - R)^{-1} \in BL(X)$ and $\|(\lambda - R)^{-1}\| \le \lambda^{-1}$ for all $\lambda > 0$. If $x \in \mathcal{D}(R)$, then $(\lambda(\lambda - R)^{-1} - 1)x = (\lambda - R)^{-1}Rx$ tends to zero as λ tends to infinity. Since $\|\lambda(\lambda - R)^{-1}\| \le 1$ for all $\lambda > 0$ and $\mathcal{D}(R)$ is dense in X, a standard argument shows that $(\lambda(\lambda - R)^{-1} - 1)x$ tends to zero as λ tends to infinity for all $x \in X$. For notational convenience in the following calculations, we let $R_\lambda = \lambda^2(\lambda - R)^{-1} - \lambda$ for all $\lambda > 0$. If $x \in \mathcal{D}(R)$, then $R_\lambda x = \lambda(\lambda - R)^{-1} Rx$ tends to Rx as λ tends to infinity.

With these little preliminaries out of the way we turn to the semigroups. Since $R_\lambda \in BL(X)$ the semigroup $t \mapsto \exp t_\lambda R : [0,\infty) \to BL(X)$ may be defined using the power series expansion for the exponential function. For each positive λ and t,

$$\begin{aligned}
&\|\exp t\, R_\lambda\| \\
&\le \exp(-t\lambda) \sum_{n=0}^{\infty} \frac{(t\lambda)^n}{n!} \|\lambda(\lambda - R)^{-1}\|^n \\
&\le 1
\end{aligned}$$

because $\|\lambda(\lambda - R)^{-1}\| \le 1$. By Lemma 6.4 R_λ and R_ν commute for all positive λ and ν. Differentiating the power series defining $\exp(wR_\lambda)$ and $\exp(wR_\nu)$ we obtain

$$\begin{aligned}
&\frac{d}{dw}\left\{\exp(wR_\lambda).\exp((t - w)R_\nu)\right\} \\
&= \exp(w\, R_\lambda).(R_\lambda - R_\nu).\exp((t - w)R_\nu)
\end{aligned}$$

for all $w, t, \lambda, \nu > 0$. Integrating this, we have

$$\begin{aligned}
&\|(\exp\, (tR_\lambda) - \exp(tR_\nu)).x\| \\
&= \left\|\int_0^t \frac{d}{dw}\left\{\exp(wR_\lambda).\ \exp((t - w)R_\nu).x\right\} dw\right\| \\
&\le \int_0^t \left\|\exp(wR_\lambda).(R_\lambda - R_\nu).\ \exp((t - w)R_\nu).x\right\| dw \\
&\le \int_0^t \|(R_\lambda - R_\nu)x\|\, dw \\
&= t\,\|(R_\lambda - R_\nu)x\|
\end{aligned}$$

for all $x \in X$. If $x \in \mathcal{D}(R)$, then $\|(R_\lambda - R_\nu)x\|$ tends to zero as λ and ν tend to infinity so that $\|(\exp(tR_\lambda) - \exp(tR_\nu)).x\|$ tends to zero uniformly in t in bounded subsets of $[0,\infty)$ as λ and ν tend to infinity. From the density of $\mathcal{D}(R)$ in X and the observation that $\|\exp(tR_\mu)\| \leq 1$ for all positive t and μ, it follows that $\|(\exp(tR_\lambda) - \exp(tR_\mu)).x\|$ tends to zero uniformly in t in bounded subse[ts] of $[0,\infty)$ as λ and ν tend to infinity for all $x \in X$. For each $t >$ [0] the operator b^t is defined by $b^t x = \lim_{\lambda\to 0} \exp(tR_\lambda).x$ for all $x \in X$. Clearly $b^t \in BL(X)$ and $\|b^t\| \leq 1$ for all $t > 0$. The uniformity of the above limits for t in bounded subsets of $[0,\infty)$ implies that $t \mapsto b^t : [0,\infty) \to BL(X)$ is strongly continuous. The semigroup property of b^t and $b^0 = I$ follow from the corresponding properties of $\exp(tR_\lambda)$.

Let $Tx = \lim_{t\to 0} t^{-1}(b^t - 1)x$ for all $x \in \mathcal{D}(T) = \{y \in X : \lim_{t\to 0} t^{-1}(b^t - 1)\, y$ exists in $X\}$. To complete the proof we show that $\mathcal{D}(T) = \mathcal{D}(R)$ and $T = R$. We do this by using a formula that occurred in Lemma 6.6. Either by integrating the power series for the exponential factor in the integrand term by term or from the proof of Lemma 6.6 (applie[d] to the semigroup $t \mapsto \exp tR_\lambda)$, we have

$$(\exp\,(t\,R_\lambda) - 1)x = \int_0^t \exp(wR_\lambda)\ R_\lambda x\ dw$$

for all $x \in X$ and all t and $\lambda > 0$. Because $\exp(w\,R_\lambda)Rx$ tends to $b^w Rx$ uniformly for $w \in [0,t]$ as λ tends to infinity, and because $\|\exp(wR_\lambda)\| \leq 1$ for all $\lambda > 0$, the above equations converge to $(b^t - 1)x = \int_0^t b^w Rx\ dw$ as λ tends to infinity for all $x \in X$. Dividin[g] by $t > 0$ and letting t tend to zero, we obtain $Tx = Rx$ for all $x \in \mathcal{D}(R)$ from the definition of T and the continuity of $w \mapsto b^w Rx : [0,\infty) \to X$. Hence $\mathcal{D}(R) \subseteq \mathcal{D}(T)$ and $R = T$ on $\mathcal{D}(R)$. If $\lambda > 0$, then $(\lambda - R)\ \mathcal{D}(R) = X$ so $(\lambda - T)\ \mathcal{D}(R) = X$, and it follows that $\mathcal{D}(R) = \mathcal{D}(T)$ because $(\lambda - T)$ is one-to-one on $\mathcal{D}(T)$. This completes the proof of the Hille-Yoshida Theorem.

6.8 EXAMPLES

We shall now sketch two examples to illustrate the above theore[m]. These examples are discussed in detail in Hille and Phillips [1974] (see

Chapter 19). Let $X = L^2(\mathbb{R})$ and define b^t on $L^2(\mathbb{R})$ by $(b^t f)(x) = f(x+t)$ for all $t \in \mathbb{R}$. Clearly b^t is an isometric operator on $L^2(\mathbb{R})$, and $t \mapsto b^t : \mathbb{R} \to BL(X)$ is a (semi)group. The strong continuity of $t \mapsto b^t$ follows easily from the density of $C_c(\mathbb{R})$, the space of continuous functions with compact support, in $L^2(\mathbb{R})$ and the strong continuity of b^t on the normed space $(C_c(\mathbb{R}), \|.\|_\infty)$. The infinitesimal generator R of the semigroup is the derivative $\frac{d}{dx}$ with $\mathcal{D}(R)$ equal to the set of $f \in L^2(\mathbb{R})$ such that $\frac{df}{dx}$ is defined almost everywhere on $\mathbb{R}$ and is in $L^2(\mathbb{R})$. That $R = \frac{d}{dx}$ on the space of continuously differentiable functions with compact support is clear. Properties of shift semigroups are intimately linked with the differential operator d/dx.

In the second example we let $X = L^1(\mathbb{R}^n)$, $b^0 f = f$, and $b^t f = G^t * f$ for all $t > 0$ and all $f \in L^1(\mathbb{R}^n)$ where G^t is the Gaussian semigroup. By Theorem 2.15 $t \mapsto b^t : [0,\infty) \to BL(X)$ is a strongly continuous contraction semigroup. The generator R of the semigroup is the Laplacian Δ with $\mathcal{D}(R)$ equal to the set of $f \in L^1(\mathbb{R}^n)$ such that Δf exists almost everywhere and $\Delta f \in L^1(\mathbb{R}^n)$. That $R = \Delta$ follows from the relation $\left(\frac{\partial}{\partial t} - \Delta\right)(G^t * f) = 0$ in Theorem 2.15.

6.9 COROLLARY

Let A be a Banach algebra. There is an analytic semigroup $t \mapsto a^t : H \to A$ such that $(a^t A)^- = A = (Aa^t)^-$ for all t in the open right half plane H and that $\|a^r\| \le 1$ for all $r > 0$ if and only if there is a $u \in A$ such that $(uA)^- = A = (Au)^-$, $\sigma(u) \cap (0,\infty) = \emptyset$, and $\|u(\lambda - u)^{-1}\| \le 1$ for all $\lambda > 0$.

Proof. Suppose that the semigroup exists. We define $\theta : L^1(\mathbb{R})^+ \to A$ by $\theta(f) = \int_0^\infty f(t)\, a^t\, dt$. Then θ is a norm reducing homomorphism from $L^1(\mathbb{R}^+)$ into A, and we may extend θ to a homomorphism from $L^1(\mathbb{R}^+)^\#$ into $A^\#$ by defining $\theta(1) = 1$. Let $v(w) = -e^{-w}$ for all $w \in \mathbb{R}^+$. Then $v^n = (-1)^n I^n$, where $I^t(w) = w^{t-1} e^{-w} \Gamma(t)^{-1}$ is the fractional integral semigroup in $L^1(\mathbb{R}^+)$. If $\lambda \in \mathbb{C}$ with $|\lambda| > 1$, then

$$(\lambda - v)^{-1}(w) = \lambda^{-1} + \lambda^{-1} \sum_{n=1}^{\infty} \frac{(-1)^n w^{n-1} e^{-w}}{\lambda^n (n-1)!}$$
$$= \lambda^{-1} - \lambda^{-1} \exp(-(1 + 1/\lambda)w)$$

for all $w \geq 0$. Since $w \mapsto \lambda^{-1} - \lambda^{-1} \exp(-(1 + 1/\lambda)w)$ is in $L^1(\mathbb{R}^+)$ for all λ in the open right half plane H, $(\lambda - v)^{-1}$ exists in $L^1(\mathbb{R}^+)^{\#}$ for all $\lambda \in H$. A similar calculation to the above shows that

$$v(\lambda - v)^{-1}(w) = \lambda^{-1} \exp(-w(1 + 1/\lambda))$$

so that $\|v(\lambda - v)^{-1}\|_1 = (1 + \lambda)^{-1}$ for all $\lambda > 0$. Further $(v*L^1(\mathbb{R}^+))^- = L^1(\mathbb{R}^+)$. We let $u = \theta(v)$. The condition $(a^t A)^- = A = (A\, a^t)^-$ implies that $(\theta(L^1(\mathbb{R}^+)).A)^- = A = (A.\theta(L^1(\mathbb{R}^+)))^-$, from which we obtain $(u\,A)^- = A = (Au)^-$. The other properties of u follow from the corresponding ones for v via the norm reducing homomorphism $\theta : L^1(\mathbb{R}^+)^{\#} \to A^{\#}$.

Conversely suppose that u exists in A with the required properties. Because $u + u^2(\lambda - u)^{-1} = \lambda u(\lambda - u)^{-1}$, we have $-u(\lambda - u)^{-1}x$ tends to ux as λ tends to zero for all $x \in A$ since $\|u(\lambda - u)^{-1}\| \leq 1$. This inequality also shows that $-u(\lambda - u)^{-1} y$ converges to y for all $y \in A$ as λ tends to zero. Similar calculations with u on the right imply that $\{-u(n^{-1} - u)^{-1} : n \in \mathbb{N}\}$ is a countable bounded approximate identity for A. Theorem 3.1 completes the proof.

6.10 NILPOTENT SEMIGROUPS

In this section we find necessary and sufficient conditions on the resolvent of the generator of a strongly continuous contraction semigroup for the semigroup to be nilpotent. The same idea is used to investigate hyperinvariant subspaces for suitable quasinilpotent operators on a Banach space.

6.11 THEOREM

Let X be a Banach space, and let $t \mapsto b^t : [0,\infty) \to BL(X)$ be a strongly continuous contraction semigroup with $b^0 = I$ and with infinitesimal generator R.

(i) The semigroup is nilpotent with $b^M = 0$ if and only if $(n!\, \|(1 - R)^{-n}\|)^{1/n} \leq M$ for all $n \in \mathbb{N}$.

(ii) There is an $M > 0$ such that $(b^M X)^-$ is neither $\{0\}$ nor X if and only if there is a non-zero $F \in X^*$ such that

$\{[n!|F((1-R)^{-n}x)|]^{1/n} : n \in \mathbb{N}\}$ is bounded for each $x \in X$.

Proof. We define $\theta : L^1(\mathbb{R}^+) \to BL(X)$ by $\theta(f)x = \int_0^\infty f(t)b^t.x \, dt$ for all $x \in X$ and all $f \in L^1(\mathbb{R}^+)$. The integral exists since $t \mapsto b^t.x : [0,\infty) \to X$ is continuous and bounded for all $x \in X$. A direct calculation shows that θ is a homomorphism from $L^1(\mathbb{R}^+)$ into $BL(X)$ such that $\|\theta\| \le 1$. Using a bounded approximate identity in $L^1(\mathbb{R}^+)$ bounded by 1 and the observation that $b^0 = I$ we obtain $\|\theta\| = 1$. From Lemma 6.6, $(1 - R)^{-1}x = \int_0^\infty e^{-w}b^w x \, dw$ for all $x \in X$ so that $(1 - R)^{-1} = \theta(I^1)$, where $I^t(w) = \Gamma(t)^{-1} w^{t-1} e^{-w}$ for w and $t > 0$ is the fractional integral semigroup in $L^1(\mathbb{R}^+)$ (see 2.6).

Suppose that $M > 0$ with $(b^M X)^- \ne X$. Using the Hahn-Banach Theorem we choose a non-zero $F \in X^*$ annihilating $b^M.X$. For each $n \in \mathbb{N}$ and each $x \in X$,

$$\begin{aligned} &|F((1 - R)^{-n}x)| \\ &= |F(\theta(I^n)x)| \\ &= \left|\int_0^\infty \frac{w^{n-1} e^{-w}}{\Gamma(n)} F(b^w x) \, dw\right| \\ &\le \int_0^M \frac{w^{n-1} e^{-w}}{\Gamma(n)} \|F\|.\|x\| \, dw \\ &\le \frac{M^n}{n!} \|F\|.\|x\| . \end{aligned}$$

This proves the necessity in (ii) and similar working, using $\|\cdots\|$ in place of $|F(\cdots)|$, proves it in case (i).

We consider the converses. In case (i) we consider all $x \in X$ with $\|x\| \le 1$ and all $F \in X^*$ with $\|F\| \le 1$, and in case (ii) we use the given $F \in X^*$ and consider all $x \in X$. Suppose that there is an $M > 0$ such that

$$|n! \, F((1 - R)^{-n}x)|^{1/n} \le M$$

for all positive integers n. In case (i) the M is independent of F and x, but in (ii) it may depend on both of them. For each $\lambda \in \mathbb{C}$

$$\begin{aligned}&\sum_{n=0}^{\infty} |F(-2\pi i\lambda)^n (1 - R)^{-n-1}x)| \\ &\le \sum_{n=0}^{\infty} \frac{(2\pi|\lambda|)^n}{(n+1)!} M^{n+1} \\ &\le M \exp (2\pi M |\lambda|),\end{aligned}$$

and hence we may define the function G by

$$G(\lambda) = \sum_{n=0}^{\infty} F((-2\pi i\lambda)^n (1 - R)^{-n-1}x)$$

for all $\lambda \in \mathbb{C}$. Then G is an entire function of exponential type $2\pi M$. If $\lambda \in \mathbb{C}$ with $\|2\pi\lambda(1 - R)^{-1}\| < 1$, then $\sum_{n=o}^{\infty} (-2\pi i\lambda)^n (1 - R)^{-n-1}$ converges in $BL(X)$ to

$$(1 - R)^{-1} (1 + 2\pi i\lambda(1 - R)^{-1})^{-1} = (1 + 2\pi i\lambda - R)^{-1}$$

so that $G(\lambda) = F((1 + 2\pi i\lambda - R)^{-1}x)$. The functions G and $\lambda \mapsto F((1 + 2\pi i\lambda - R)^{-1}x)$ are analytic in a neighbourhood of the closed lower half plane $-iH^-$, and hence these functions are equal on $-iH^-$. By the Hille-Yoshida Theorem

$$(1 + 2\pi i\lambda - R)^{-1}x = \int_0^{\infty} e^{-(1+2\pi i\lambda)w}\, b^w\, x\, dw$$

for each $\lambda \in \mathbb{R}$, and so

$$G(\lambda) = \int_0^{\infty} e^{-2\pi i\lambda w} F(e^{-w}b^w x)\, dw.$$

We define the function K on $\mathbb{R}$ by $K(w) = F(e^{-w}b^w x)$ for $w \ge 0$ and $K(w) = 0$ for $w < 0$. Then $|K(w)| \le e^{-w}$ for all $w > 0$ and $K \in L^1(\mathbb{R}) \cap L^2(\mathbb{R})$. Further G is the Fourier transform $K^\wedge$ of K, and by Plancherel's Theorem the restriction of G to $\mathbb{R}$ is in $L^2(\mathbb{R})$ (see Rudin [1966]). Because G has exponential type $2\pi M$ the Paley-Wiener Theorem (see Rudin [1966]) implies that the support of K is contained in $[-M,M]$. Note that the 2π occurring in the exponent of our definition of the Fourier transform appears in the relationship between the support of

K and the type of G. Thus $F(b^w x) = 0$ for all $w \geq M$. In case (i) we are done because M was independent of F and x — the Hahn-Banach Theorem ensures that $b^M = 0$. In case (ii) we let $B_j = \{x \in X : F(b^w x) = 0$ for all $w \geq j\}$ for each $j \in \mathbb{N}$. Then each B_j is a closed linear subspace of X, and what we have just proved shows that $\bigcup_{j=1}^{\infty} B_j = X$. The Baire Category Theorem implies that $B_N = X$ for some positive integer N, and so $F(b^N X) = \{0\}$ and $(b^N X)^- \neq X$. However $b^N X$ could be $\{0\}$. If $0 < t < r$, then $(b^t X)^- \supseteq (b^r X)^-$. Also if $(b^r X)^- = X$ for some $r > 0$, then $(b^t X)^- = X$ for all $t > 0$ since $(b^{r+s} X)^- = (b^s (b^r X)^-)^- = (b^s X)^-$. Thus $(b^t X)^- \neq X$ for all $t > 0$, and the strong continuity of $t \mapsto b^t : [0,\infty) \to BL(X)$ and $b^0 = I$ ensure that $(b^t X)^-$ is nonzero for small enough t. This completes the proof.

6.12 REMARKS

Note that by using Stirling's Formula,

$$n\Gamma(n) = (2\pi)^{1/2} n^{1/2} n^n e^{-n+O(1/n)}$$

as n tends to infinity, the condition $\|n!\,(1 - R)^{-n}\|^{1/n} \leq M$ in Theorem 6.11(i) may be replaced by $\{n\,\|(1 - R)^{-n}\|^{1/n} : n \in \mathbb{N}\}$ is bounded, but with the loss of the nice relationship between $b^M = 0$ and the bound M.

If T is a continuous linear operator on a Banach space X, then a hyperinvariant subspace Y for T is a proper closed linear subspace Y of X such that $SY \subseteq Y$ for all $S \in BL(X)$ commuting with T. The second commutant of an operator T on a Banach space X is the set of all $R \in BL(X)$ such that R commutes with all $S \in BL(X)$ commuting with T.

6.13 THEOREM

Let T be a non-zero continuous linear operator on a Banach space X satisfying $(0,\infty) \cap \sigma(T) = \emptyset$, and $\|T(\lambda - t)^{-1}\| \leq 1$ for all $\lambda > 0$.

((i) If $(TX)^- = X$ and $\{n\,\|T^n\|^{1/n} : n \in \mathbb{N}\}$ is bounded, then there is a nilpotent strongly continuous semigroup $t \mapsto b^t : [0,\infty) \to BL(X)$ with $b^0 = I$ and b^t in the second commutant of T for all $t > 0$.

(ii) If there is a non-zero $F \in X^*$ such that $\{n\,|F(T^n x)|^{1/n} : n \in \mathbb{N}\}$

is bounded for all $x \in X$, then T has a hyperinvariant subspace.

Proof. (i) From the equation $T + T^2(\lambda - T)^{-1} = \lambda T(\lambda - T)^{-1}$ and the inequality $\|T(\lambda - T)^{-1}\| \le 1$ for all $\lambda > 0$, it follows that $-T(\lambda - T)^{-1}Tx$ tends to Tx as λ tends to zero. Hence $-T(\lambda - T)^{-1}y$ tends to y as λ tends to zero for all $y \in X$ because $(TX)^- = X$. Therefore T is a one-to-one continuous linear operator from X onto the dense linear subspace TX of X, and so $R = T^{-1} : \mathcal{D}(R) = TX \to X$ is a closed operator satisfying $T(\lambda - T)^{-1} = (\lambda R - 1)^{-1} \in BL(X)$ and $\|(\lambda R - 1)^{-1}\| \le 1$ for all $\lambda > 0$ (see 6.3). Further $\|(1 - R)^{-n}\|^{1/n} \le \|T^n\|^{1/n}\,\|1 - T\|$ for all $n \in \mathbb{N}$ so that R satisfies the hypotheses of Theorem 6.11(i) as modified by Stirling's Formula in Remark 6.12. Hence the strongly continuous semigroup $t \mapsto b^t : [0,\infty) \mapsto BL(X)$ generated by R is nilpotent. From the Hille-Yoshida Theorem b^t is the strong limit of $\exp t(\lambda^2(\lambda - R)^{-1} - \lambda) = \exp t(\lambda^2\, T(\lambda T - 1)^{-1} - \lambda)$ as λ tends to infini
Because the second commutant of T is closed under strong limits, b^t is in the second commutant of T for all $t > 0$. The strong continuity of $t \mapsto b^t : [0,\infty) \to BL(X)$, $b^0 = I$, and $b^N = 0$ for some positive N ensure that there is an $r > 0$ such that $(b^r X)^-$ is neither $\{0\}$ nor X. Since b^r is in the second commutant of t, $(b^r X)^-$ is a hyperinvariant subspace of X.

(ii). We may suppose that $(TX)^- = X$, for otherwise $(TX)^-$ is a hyper-invariant subspace of X. By the first part of the proof of (i), $R = T^{-1}$ is the infinitesimal generator of a strongly continuous semigroup $t \mapsto b^t : [0,\infty) \to BL(X)$ in the second commutant of T. By Stirling's Formula applied as in Remark 6.12, the boundedness of the set $\{n|F(T^n x)|^{1/n} : n \in \mathbb{N}\}$ for each $x \in X$ is equivalent to the boundedness of the set $\{(n!|F(T^n x)|)^{1/n} : n \in \mathbb{N}\}$ for each $x \in X$. Suppose that $n!\,|F(T^n x)| \le M^n$ for all $n \in \mathbb{N}$ where M depends on x. Then

$$\begin{aligned}
& n!\,|F(1 - R)^{-n}x)| \\
&= n!\,|F(T^n(T - 1)^{-n}x)| \\
&\le n! \sum_{j=0}^{\infty} \left|\binom{-n}{j}\right|\,|F(T^{n+j}x)| \\
&\le n! \sum_{j=0}^{\infty} \frac{(n + j - 1)(n + j - 2)\cdots n.M^{n+j}}{j!\,(n + j)!} \\
&\le M^n e^M
\end{aligned}$$

for all $n \in \mathbb{N}$. By Theorem 6.1 it follows that $(b^r X)^-$ is neither $\{0\}$ nor X for some $r > 0$. This completes the proof.

6.14 PROBLEM

Can the hypothesis $\|T(\lambda - T)^{-1}\| \le 1$ for all $\lambda > 0$ be omitted from the hypotheses of Theorem 6.13?

6.15 EXAMPLE

Theorem 6.13 may be used to obtain the obvious hyperinvariant subspaces of the Volterra operator $T : f \mapsto \int_0^x f(w)\,dw : L^2[0,1] \to L^2[0,1]$ by showing that T satisfies the hypotheses of Theorem 6.13. The strongly continuous semigroup $t \mapsto b^t : [0,\infty) \to BL(L^2[0,1])$ given by the theorem is the shift semigroup on $L^2[0,1]$ defined by

$$(b^t f)(x) = \begin{cases} 0 & 0 \le x < t \\ f(t - x) & t \le x \le 1 \end{cases}$$

for all $t \ge 0$ and $f \in L^2[0,1]$. However what is really happening in this example is that the closed hyperinvariant subspaces arise from $L^2[0,1]$ being a Banach module over $L^1_*[0,1]$. We shall discuss this in more detail later in the chapter, but we illustrate it briefly here. We have a norm reducing homomorphism θ from the Volterra algebra $L^1{}_*[0,1]$ into $BL(L^2[0,1])$ given by $\theta(f)g = f*g$ for all $f \in L^1[0,1]$ and $g \in L^2[0,1]$, and $\theta(u) = T$ where $u(w) = 1$ for all $w \in [0,1]$. The properties of u in $L^1_*[0,1]$ give the hypotheses of Theorem 6.13(i) for T except for the condition $(TX)^- = X$ which must be checked directly.

6.16 PROPER CLOSED IDEALS IN RADICAL BANACH ALGEBRAS

We now turn to continuous semigroups in a Banach algebra, and relate these to proper closed ideals in the algebra. The main problem in this area of research is, does a commutative radical Banach algebra A have a proper closed ideal? If there is a non-zero $b \in A$ such that $x \mapsto bx : A \to A$ is a compact linear operator, then Lomonosov's Theorem (see Radjavi and Rosenthal [1973]) implies that this operator has a hyperinvariant subspace, and so A has a proper closed ideal. In a similar way we shall convert the operator theory results of the previous section into Banach algebra results. Before turning to our proper closed ideals and nilpotent semigroups, we briefly consider quasinilpotent semigroups.

6.17 THEOREM

Let A be a Banach algebra, let $t \mapsto a^t : (0,\infty) \to A$ be a continuous contraction semigroup, and let $u = \int_0^\infty e^{-t} a^t \, dt$ be in A. The semigroup $t \mapsto a^t$ is quasinilpotent if and only if u is quasinilpotent.

Proof. Suppose a^1 is quasinilpotent. If $t > 1$, then $0 \le \|(a^t)^n\|^{1/n} \le \|a^n\|^{1/n}$, since $tn - n > 0$, so the spectral radius of a^t is zero. If $t > 0$, there is an $m \in \mathbb{N}$ such that $mt > 1$ thus a^{mt} and a^t are quasinilpotent. A commutative Banach algebra generated by quasinilpotent elements is a radical algebra. Hence u is quasinilpotent as it is in the commutative Banach algebra generated by $\{a^t : t > 0\}$.

Conversely let B be the commutative Banach algebra generated by $\{a^t : t > 0\}$, and suppose that B is not a radical algebra. Then there is a character ϕ on B, and $t \mapsto \phi(a^t) : [0,\infty) \to \mathbb{C}$ is a continuous semigroup so there is a $\beta \in \mathbb{C}$ such that $\phi(a^t) = e^{\beta t}$ for all $t > 0$. Since $t \mapsto a^t$ is a contraction semigroup $\mathrm{Re}\,(\beta) \le 0$. Hence $\phi(u) = \int_0^\infty e^{-t} e^{\beta t} \, dt = (-1 + \beta)^{-1}$ is non-zero, contrary to the quasinilpotence of u.

6.18 PROBLEM

Let A be a Banach algebra, and let $t \mapsto a^t : (0,\infty) \to A$ be a continuous contraction semigroup such that $(\bigcup_{t>0} a^t A)^- = A = (\bigcup_{t>0} A a^t)^-$. If the semigroup is quasinilpotent, is A a radical Banach algebra?

6.19 THEOREM

Let A be a commutative Banach algebra. There is a continuous contraction nilpotent semigroup $t \mapsto a^t : (0,\infty) \to A$ with $(\bigcup_{t>0} a^t A)^- = A$ if and only if there is a non-zero $u \in A$ such that $(uA)^- = A$, $\sigma(u) \cap (-\infty,0) = \emptyset$, $\|u(\lambda + u)^{-1}\| \le 1$ for all $\lambda > 0$, and $\{n\|u^n\|^{1/n} : n \in \mathbb{N}\}$ is bounded.

Proof. To prove this result we shall combine the operator theory results of this chapter with the existence of a suitable continuous semigroup given by Theorem 3.1. Suppose that a continuous nilpotent contraction semigroup $t \mapsto a^t : (0,\infty) \to A$ exists satisfying $(\bigcup_{t>0} a^t A)^- = A$. We define

$b^t \in BL(A)$ by $b^t(x) = a^t.x$ for all $t > 0$ and $x \in A$, and let $b^0 = I$. Then $t \mapsto b^t : [0,\infty) \to BL(A)$ is a contraction semigroup, which is strongly continuous since $(\bigcup_{t>0} a^t A)^- = A$. Let R be the infinitesimal generator of this semigroup, and let $u = \int_0^\infty e^{-t} a^t \, dt$. Note that this integral converges in A because $t \mapsto e^{-t}a^t : (0,\infty) \to A$ is continuous and $\|e^{-t}a^t\| < e^{-t}$. From the Hille-Yoshida Theorem we see that $(1 - R)^{-1}.x = ux$ for all $x \in A$, and so the operator $(1 - R)^{-1}$ is left multiplication L_u by u. Hence

$$(\lambda + L_u)^{-1} = (\lambda + (1 - R)^{-1})^{-1} = (1 - R)^{-1}((\lambda + 1)\lambda^{-1} - R)^{-1} \lambda^{-1}$$

is in $BL(A)$ and

$$L_u(\lambda + L_u)^{-1} = (1 - R)^{-1}(\lambda + (1 - R)^{-1})^{-1} = \lambda^{-1} ((1 + \lambda)\lambda^{-1} - R)^{-1}$$

for all $\lambda > 0$. Since the spectra of u in A and L_u in $BL(A)$ are equal, $\sigma(u) \cap (-\infty,0) = \emptyset$. Further

$$\|u(\lambda + u)^{-1}\| = \|L_u(\lambda + L_u)^{-1}\|$$

because A has a bounded approximate identity $\{a^{1/n} : n \in \mathbb{N}\}$, say, bounded by 1. The estimate on $\|(\mu - R)^{-1}\|$ in the Hille-Yoshida Theorem gives

$$\begin{aligned} \|u(\lambda + u)^{-1}\| &= \lambda^{-1}\|((1 + \lambda)\lambda^{-1} - R)^{-1}\| \\ &\le \lambda^{-1} ((1 + \lambda)\lambda^{-1})^{-1} = (\lambda + 1)^{-1} \le 1 \end{aligned}$$

for all $\lambda > 0$. Also $uA = (1 - R)^{-1}A = \mathcal{D}(R)$ is dense in A. The semigroup b^t is nilpotent, because a^t is nilpotent, and thus the set $\{n\|(1 - R)^{-n}\|^{1/n} : n \in \mathbb{N}\}$ is bounded by Theorem 6.11 and Remark 6.12. This completes the proof of this implication because $(1 - R)^{-n} = u^n$ for all $n \in \mathbb{N}$.

Conversely suppose that there is a u in A with the required properties. We define $T : x \mapsto -ux : A \to A$. Then T satisfies the hypotheses of Theorem 6.13, and there is a strongly continuous contraction semigroup $t \mapsto b^t : [0,\infty) \to BL(A)$ with $b^0 = I$ and $b^N = 0$. Since $u - u^2(\lambda + u)^{-1} = \lambda u(\lambda + u)^{-1}$ for $\lambda > 0$, we have $u(\lambda + u)^{-1}ux$ tends to ux as λ tends to zero for each $x \in A$. Because $\|u(\lambda + u)^{-1}\| \le 1$ and

$(uA)^- = A$, it follows that $u(\lambda + u)^{-1}y$ tends to y for each $y \in A$ as λ tends to zero. Thus $\{u(n^{-1} + u)^{-1} : n \in \mathbb{N}\}$ is a countable bounded approximate identity in A bounded by 1. By Theorem 3.1 there is a continuous contraction semigroup $t \mapsto a^t : (0,\infty) \to A$ such that $(a^t A)^- = A$ for all $t > 0$. The map $t \mapsto b^t(a^t) : (0,\infty) \to A$ is a continuous contraction semigroup, because $t \mapsto a^t : (0,\infty) \to A$ is a continuous contraction semigroup and $t \mapsto b^t$ is a strongly continuous contraction semigroup into the multiplier algebra of A. Finally $(b^t a^t A)^- = (b^t A)^-$ for all $t > 0$, and so $t \mapsto b^t a^t : (0,\infty) \to A$ is the required nilpotent semigroup.

Theorem 6.19 may be used to give conditions on a Banach algebra that ensure that there is a continuous homomorphism from $L^1_*[0,1]$ into the Banach algebra.

6.20 COROLLARY

Let A be a commutative Banach algebra. There is a continuous norm reducing monomorphism θ from the Volterra algebra $L^1_*[0,1]$ into A such that $(\theta(L^1_*[0,1]).A)^- = A$ if and only if there is $u \in A$ such that $(uA)^- = A$, $\|u(\lambda - u)^{-1}\| \leq 1$ for all $\lambda > 0$, and $\{n\|u^n\|^{1/n} : n \in \mathbb{N}\}$ is bounded.

Proof. If a u with these properties exists in A, then by Theorem 6.19 there is a continuous contraction semigroup $t \mapsto a^t : (0,\infty) \to A$ such that $(\bigcup_{t>0} a^t A)^- = A$ and $a^N = 0$ for some N. By a change of scale in t we may assume that $a^t = 0$ if and only if $t \geq 1$. We let $\theta : f \mapsto \int_0^1 f(t)\, a^t\, dt : L^1_*[0,1] \to A$. Clearly θ is a norm reducing homomorphism from $L^1_*[0,1]$ into A, and the property $(\theta(L^1_*[0,1]).A)^-$ follows from $(\bigcup_{t>0} a^t A)^- = A$. If θ is not one-to-one, then the kernel of θ is a proper closed ideal J in $L^1_*[0,1]$. Each closed ideal J in the Volterra algebra is of the form $\{f \in L^1_*[0,1] : f = 0$ a.e. on $[0,\alpha]\}$ for some $\alpha \geq 0$ (see Dales [1978], Dixmier [1949], or Radjavi and Rosenthal [1973]). Since J is assumed to be non-zero, the α corresponding to J satisfies $\alpha < 1$. If f_n is the characteristic function of the interval $[\alpha, \alpha + 1/n]$, then $f_n \in J$ and $n\,\theta(f_n)$ tends to a^α as n tends to infinity. Thus $a^\alpha = 0$, which gives a contradiction so θ is one-to-one.

Conversely if the norm reducing monomorphism exists, we let $u = \theta(v)$ where $v(t) = 1$ for $t \in [0,1]$. The properties of u now follow from the corresponding ones for v and from $(\theta(L^1_*[0,1]).A)^- = A$.

In the proof of Theorem 6.19 we used Theorem 6.13(i) with other techniques. By modifying the proof of Theorem 6.19 slightly and using Theorem 6.13(ii) we obtain the following Theorem, whose proof we omit.

6.21 THEOREM

Let A be a commutative Banach algebra. There is a continuous contraction semigroup $t \mapsto a^t : (0,\infty) \to A$ such that $(\bigcup_{t>0} a^t A)^- = A$ and $(a^r A)^-$ is neither A nor $\{0\}$ for some $r > 0$ if and only if there is a non-zero $F \in A^*$ and $u \in A$ with $(uA)^- = A$, $\sigma(u) \cap (-\infty,0) = \emptyset$, $\|u(\lambda + u)^{-1}\| \le 1$ for all $\lambda > 0$, and $\{n|F(u^n x)|^{1/n} : n \in \mathbb{N}\}$ is bounded for all $x \in A$.

6.22 EXAMPLES

We shall now consider two examples of Banach algebras that satisfy Theorems 6.19 and 6.21.

In the Volterra algebra $L^1_*[0,1]$ we could take $u(s) = 1$ for all $s \in [0,1]$. The strongly continuous semigroup in the multiplier algebra $\mathrm{Mul}(L^1_*[0,1])$ of $L^1_*[0,1]$ corresponding to u is $t \mapsto \delta^t$, where δ^t is the unit point mass at t for $0 \le t \le 1$ and is zero for $t > 1$. A corresponding continuous contraction semigroup in $L^1_*[0,1]$ could be $t \mapsto \delta^t * I^t$, where $t \mapsto I^t$ is the fractional integral semigroup of $L^1(\mathbb{R}^+)$ restricted as a function on $\mathbb{R}^+$ to $[0,1]$.

Let $\mathfrak{A}$ be the Banach algebra of continuous functions in the closed right half plane H^-, analytic in H, and tending to zero as $|z|$ tends to infinity. This algebra was introduced in Example 5.17. Let u in $\mathfrak{A}$ be defined by $u : z \mapsto (z + 1)^{-1} : H^- \to \mathbb{C}$. Then u and $(\lambda + u)^{-1}$ are in $\mathfrak{A}$ for all $\lambda > 0$, and

$$u(\lambda + u)^{-1}(z) = (1 + \lambda + \lambda z)^{-1}$$

for all $z \in H^-$. Thus $\|u(\lambda + u)^{-1}\| \le 1$ for all $\lambda > 0$. The strongly continuous contraction semigroup of multipliers generated by $R : uf \mapsto -f : u\mathfrak{A} \to \mathfrak{A}$ is $b^t(z) = e^{-t}.e^{-tz}$ for all $z \in H^-$ and $t > 0$. Further $u \in (b^t\mathfrak{A})^-$ for all $t > 0$.

6.23 PROBLEM

Is there a nice class of Banach algebras with countable bounded approximate identities such that we can find all proper closed ideals in each algebra? If there is a continuous norm reducing monomorphism θ from $L^1_*[0,1]$ into a commutative Banach algebra A such that $(\theta(L^1_*[0,1]).A)^- = A$, what additional properties are required to ensure that all proper closed ideals in A arise from proper closed ideals in $L^1_*[0,1]$?

6.24 REMARKS AND NOTES

Our discussion of the Hille-Yoshida Theorem is standard and as we have noted earlier there are accounts in several references (see Hille and Phillips [1974], Dunford and Schwartz [1958], and Reed and Simon [1975]). There is a nice account of the generators of analytic semigroups in Reed and Simon [1975]. From Theorem 6.11 onwards the results in this chapter are taken from the seminars and postgraduate lectures that Jean Esterle gave at U.C.L.A. in 1979. Though the discussion differs slightly from his, these results are due to Esterle. At present Esterle has not published these results, and I am grateful for his permission to include them in these notes.

APPENDIX 1 : THE AHLFORS-HEINS THEOREM

In this appendix we shall prove a special case of the Ahlfors-Heins Theorem which will be strong enough for the applications in these notes. Our hypotheses are stronger than those of the full result in that we require $\int_{\mathbb{R}} \frac{\log^+|f(iy)|dy}{1+y^2} < \infty$ rather than $\int_1^\infty y^{-2} \log^+|f(iy)\ f(-iy)\ dy| < \infty$ and analyticity in a neighbourhood of the closed half plane. For a discussion of this point see Boas [1954, p.114]. The conclusions in the theorem we prove are weaker in that the convergence holds except for a set of measure zero rather than except in a set of outer capacity zero. Our proof will assume the complex analysis that is in Rudin [1966], Real and Complex Analysis, and so on the way we shall prove results of Carleman and Nevanlinna on analytic functions of exponential type in a half plane.

We define $\log^+ w = 0$ if $0 < w < 1$ and $\log^+ w = \log w$ if $w \geq 1$. Further $\log^- w = \log w - \log^+ w$ for all $w > 0$. Note that $\log^+$ is an increasing function of w satisfying $\log^+ (uw) \leq \log^+ u + \log^+ w$ for all $u, w > 0$. Recall that a function f analytic in the open right half plane is of *exponential type*, or of *exponential type τ*, if there is a non-negative real number τ such that

$$\limsup_{R \to \infty} R^{-1} \log (\sup\{|f(z)| : z \in H, |z| \leq R\}) \leq \tau.$$

If f is of exponential type τ, then for each $\varepsilon > 0$ there is a constant C such that $|f(z)| \leq C \exp((\tau + \varepsilon)|z|)$ for all $z \in H$.

A1.1 THEOREM (AHLFORS-HEINS)

Let f be analytic in a neighbourhood of the closed right half plane H^-, and let f be of exponential type in H. If $\int_{\mathbb{R}} \frac{\log^+|f(iy)|dy}{1+y^2}$ is finite, then

$$\alpha = \lim_{r\to\infty} \frac{2}{\pi r}\int_{-\pi/2}^{\pi/2} \cos\theta \log|f(re^{i\theta})|\, d\theta$$

exists in $\mathbb{R}$, and

$$\lim_{r\to\infty} r^{-1} \log|f(re^{i\theta})| = \alpha\cos\theta$$

for almost all $\theta \in (-\pi/2, \pi/2)$.

We shall firstly prove a classical theorem of Carleman on functions analytic in a half plane, and obtain a number of properties of f from this theorem. These properties and the Poisson formula for a semi-disc are combined to give a proof of the theorem of Nevanlinna which expresses $\log|f(z)|$ in terms of the zeros of f in H and in terms of a convolution with a Poisson kernel along $i\mathbb{R}$. In the proof of the Ahlfor Heins Theorem it is only the logarithm of the Blaschke product of the zerc of f that requires a rather delicate argument. The proofs are broken up into lemmas, and our first lemma helps in various calculations with the zeros of f.

A1.2 LEMMA

If $\beta \in H$ with $|\beta| < R$ and if

$$h(z) = \left(\frac{1 - z/\beta}{1 + z/\bar\beta}\right)\cdot\left(\frac{R^2 + \beta z}{R^2 - \bar\beta z}\right)$$

for all $z \in H$, then $|h(z)| = 1$ for all $z \in i\mathbb{R} \cup \{z \in \mathbb{C} : |z| = R\}$ an $h'(0) = 2(\text{Re}\,\beta)\left(\frac{1}{R^2} - \frac{1}{|\beta|^2}\right)$.

Proof. Expanding h in a power series in z we obtain

$$h(z) = 1 + z\left(-\frac{1}{\beta} - \frac{1}{\bar\beta} + \frac{\beta}{R^2} + \frac{\bar\beta}{R^2}\right) + z^2\ldots,$$ which gives $h'(0)$

Now

$$h(iy) = \left(\frac{1 - iy/\beta}{1 - \overline{iy/\beta}}\right)\cdot\left(\frac{R^2 + iy\beta}{R^2 + \overline{iy\beta}}\right)$$

for all $y \in \mathbb{R}$, and $h(Re^{i\theta}) = \frac{\bar{\beta}}{\beta}\left(\frac{\beta - Re^{i\theta}}{\bar{\beta} + Re^{i\theta}}\right) \cdot \left(\frac{Re^{-i\theta} + \beta}{Re^{-i\theta} - \bar{\beta}}\right)$ for all $\theta \in \mathbb{R}$.

The lemma follows from these equalities because $|\beta| = |\bar{\beta}|$.

A1.3 THEOREM (Carleman's Theorem)

Let f be analytic in a neighbourhood of the closed right half plane H^- with $f(0) = 1$, and let $z_k = r_k . e^{i\theta_k}$ be the zeros of f in the open right half plane H, repeated according to their multiplicities. If $R > 0$ and R is not an r_k, then

$$\sum_{r_k<R}\left(\frac{1}{r_k^{\,2}} - \frac{1}{R^2}\right) r_k \cos \theta_k$$

$$= \frac{1}{\pi R} \int_{-\pi/2}^{\pi/2} \cos \psi \log |f(Re^{i\psi})| \, d\psi$$

$$+ \frac{1}{2\pi} \int_0^R \left(\frac{1}{y^2} - \frac{1}{R^2}\right) \log |f(iy).f(-iy)| \, dy - \frac{1}{2} \operatorname{Re} f'(0).$$

The last integral is convergent.

Proof. Firstly suppose that f has no zeros in $\{z \in H : |z| \le R\}$. Let ρ and r be small and positive. Let Γ be the positively oriented closed contour which follows the semicircles $|z| = \rho$ and $|z| = R$ in H and the y-axis between iR and $-iR$ except that where there is a zero of f on the y-axis the contour detours round it by a small semicircle with centre the zero and radius r in H. By Theorem 13.18 Rudin [1966, p.262] since $\{z \in H : |z| < R\}$ is simply connected and $f(0) = 1$, we may choose a logarithm function log such that $z \mapsto \log f(z)$ is analytic in this set and $\log f(z) = f'(0) z + \ldots$ for z near zero in H.

The integral

$$I_{\rho,r} = \frac{1}{2\pi i} \int_\Gamma \left(\frac{1}{z^2} + \frac{1}{R^2}\right) \log f(z) \, dz$$

is zero, because the integrand is analytic in a neighbourhood of the region surrounded by Γ. For ρ very small the integral round the semicircle of radius ρ is approximately

$$\frac{1}{2\pi i}\int_{\pi/2}^{-\pi/2} f'(0)\,\rho e^{i\psi}\left(\frac{1}{\rho^2 e^{2i\psi}}+\frac{1}{R^2}\right) i\rho e^{i\psi}\,d\psi = -\,f'(0)/2.$$

The integral round the semicircle of radius R is

$$\frac{1}{2\pi i}\int_{-\pi/2}^{\pi/2} (R^{-2}\,e^{-2i\psi}+R^{-2})\;i\;R\;e^{i\psi}\,\log(f(Re^{i\psi}))\,d\psi$$

$$=\frac{1}{\pi R}\int_{-\pi/2}^{\pi/2}\cos\psi\,\log f(Re^{i\psi})\,d\psi.$$

Near a zero α of order m of f on the y-axis, $|(z^{-2}+R^{-2})\log f(z)|$ is bounded above by $C + K\,|\log|z-\alpha||$ where C and K are constants depending on α, m, and the determination of the logarithm near $f(\alpha)$. Because $r\,|\log r| \to 0$ as $r \to 0$, the integral round each little semicir detouring a zero of f on the y-axis tends to zero as r tends to zero. After taking the limit as r tends to zero the integral down the y-axis i

$$-\frac{1}{2\pi i}\int_{R\geq|y|\geq\rho}\{-y^{-2}+R^{-2}\}i\,\log f(iy)\,dy$$

$$=-\frac{1}{2\pi}\int_{\rho}^{R}\{R^{-2}-y^{-2}\}\,\log(f(iy)\,f(-iy))\,dy$$

$$\to\frac{1}{2\pi}\int_{0}^{R}\{y^{-2}-R^{-2}\}\,\log\,(f(iy)\,f(-iy))\,dy$$

as $\rho \to 0$. Now taking the real part of the limit of $I_{\rho,r}$ as $r \to 0$ then $\rho \to 0$ we obtain zero is equal to the right hand side of Carleman's equation, that is,

$$0=\frac{1}{\pi R}\int_{-\pi/2}^{\pi/2}\cos\psi.\log|f(Re^{i\psi})|\,d\psi$$

$$+\frac{1}{2\pi}\int_{0}^{R}(R^{-2}-y^{-2})\,\log|f(iy)\,f(-iy)|\,dy-\frac{1}{2}\,\mathrm{Re}\,f'(0)$$

Finally suppose that $z_1,\ldots,z_n$ are the zeros of f in $\{z \in H : |z| < R\}$ with each zero repeated according to its multiplicity, and let $f(z) = g(z)\,h(z)$, where

$$h(z)=\prod_{j=1}^{n}\left(\frac{1-z/z_j}{1+z/\bar{z}_j}\right)\left(\frac{R^2+z\bar{z}_j}{R^2-zz_j}\right)$$

Then $g(z)$ is analytic in a neighbourhood of $\{z \in H^- : |z| \leq R\}$, g has no zeros in $\{z \in H : |z| < R\}$, and $g(0) = 1$. Further $|h(z)| = 1$ for all $z \in i\mathbb{R} \cup \{w \in \mathbb{C} : |w| = R\}$ by Lemma A1.2, so that the result will follow once we have proved that $1/_2 \operatorname{Re} h'(0)$ is the left hand side of the equation in the statement (because $h(0) = g(0) = 1$). Since each factor in h has value 1 at zero, from Lemma A1.2 we obtain

$$1/_2\, h'(0) = \sum_1^n (R^{-2} - z_j^{-2}) \operatorname{Re} z_j$$

as required.

A1.4 COROLLARY

Let f be analytic in a neighbourhood of H^- with $f(0) \neq 0$, and let f be of exponential type in H with zeros $z_1, z_2, \ldots$ in H repeated according to their multiplicities. If $\int_{\mathbb{R}} \frac{\log^+ |f(iy)|}{1 + y^2}\, dy$ is finite, then

(i) $\sum \operatorname{Re} (z_n^{-1})$ converges,

(ii) $\sum \left|\log \left|\frac{1 - z/z_n}{1 + \bar{z}/z_n}\right|\right|$ converges for all $z \in H$ not a zero of f,

(iii) $\int_{\mathbb{R}} \frac{|\log |f(iy)||}{1 + y^2}\, dy$ is finite, and the set

(iv) $\left\{\frac{1}{\pi R} \int_{-\pi/2}^{\pi/2} \cos \psi |\log | f(Re^{i\theta})||\, d\psi : R \geq 1\right\}$ is bounded.

Proof. By dividing f by a constant we may assume that $f(0) = 1$. Let $z_n = r_n e^{i\theta_n}$ for all n, and let M and k be constants such that $|f(z)| \leq e^{M|z|+k}$ for all $z \in H$. Choose $r > 0$ such that $r < r_n$ for all n. If R is large and not equal to an r_n, then by Carleman's Theorem

$$\sum_{r_n < R} (r_n^{-2} - R^{-2})\, r_n \cos \theta_n$$

$$(1)\quad = \frac{1}{\pi R}\int_{-\pi/2}^{\pi/2} \cos\psi . \log|f(Re^{i\psi})|d\psi$$

$$+ \frac{1}{2\pi}\int_{0}^{R} (y^{-2} - R^{-2}) \log|f(iy). f(-iy)| dy - 1/_2 \operatorname{Re} f'(0)$$

$$\le \frac{\pi}{\pi R} (MR + k) + \frac{1}{2\pi}\int_{r}^{R} (y^{-2} - R^{-2}) \log^+ |f(iy) f(-iy)|dy + C,$$

where C is a constant depending on r and f but independent of R. The last integral is bounded above by $\frac{1}{2\pi}\int_{r\le|y|} y^{-2}\log^+|f(iy)|\, dy$, which is finite. Hence there is a β such that the increasing function

$$R \mapsto \sum_{r_n<R} (r_n^{-2} - R^{-2}) r_n \cos\theta_n$$

converges to β as R tends to infinity through positive real values not equal to any r_n. Thus $\sum \frac{\cos\theta_n}{r_n} = \sum \operatorname{Re}(z_n^{-1})$ converges because its partial sums are bounded above by β. This proves (i).

If $w \in \mathbb{C}$ with $|w| \le 1/2$, then $|\log|1 + w|| \le 2|w|$. Now

$$\frac{1 - z/z_n}{1 + z/\bar{z}_n} = 1 - \frac{2z \operatorname{Re}(z_n^{-1})}{1 + z/\bar{z}_n}$$

so that

$$(2)\quad \left|\log\left|\frac{1 - z/z_n}{1 + z/\bar{z}_n}\right|\right| \le 4|z| \frac{\operatorname{Re}(z_n^{-1})}{|1 + z/\bar{z}_n|} \le 8|z| \operatorname{Re}(z_n^{-1})$$

for $z \in \mathbb{C}$ with $|z| \operatorname{Re}(z_n^{-1}) < 1/_2$. Hence if $z \in H$ is not a zero of f, then $\sum_n \left|\log\left|\frac{1 - z/z_n}{1 + z/\bar{z}_n}\right|\right|$ converges. This is (ii). Note that since $\left|\frac{1 - z/z_n}{1 + z/\bar{z}_n}\right|$ is less than 1 for $z \in H$ its logarithm is negative.

If R is large and R is not an r_n, then using $\log = \log^+ + \log^-$ and part of inequality (1) we obtain

$$- \frac{1}{2\pi} \int_{r \le |y| \le R} (y^{-2} - R^{-2}) \log^- |f(iy)| \, dy$$

$$\le C + R^{-1}(MR + k) + \sum_{r_j < R} \frac{\cos \theta}{r_j} + \frac{1}{2\pi} \int_{r \le |y| \le R} \frac{\log^+ |f(iy)|}{y^2} \, dy.$$

The right hand side of this inequality is bounded above for all specified R since the series and integral converge as R tends to infinity. If $r \le |y| \le R/2$, then $(y^{-2} - R^{-2}) \ge \frac{3}{4} y^{-2}$ so that $- \frac{3}{4} \int_{r \le |\ | \le R} y^{-2} \log^- |f(iy)| dy$ is bounded above for all r. It follows that $\int_{\mathbb{R}} - \frac{\log^- |f(iy)|}{1 + y^2} dy$ is finite, and this implies the convergence of $\int_{\mathbb{R}} \frac{|\log |f(iy)||}{1 + y^2} dy$ because $|\log| = \log^+ - \log^-$. This proves (iii).

From Carleman's Theorem for R not the modulus of a zero of f, we have

$$- \frac{1}{\pi r} \int_{-\pi/2}^{\pi/2} \cos \psi \log^- |f(R e^{i\psi})| \, d\psi$$

$$\le \sum_{j=1}^{\infty} \frac{\cos \theta_j}{r_j} + \frac{1}{\pi R} \int_{-\pi/2}^{\pi/2} \cos \psi \, |f(R e^{i\psi})| \, d\psi$$

$$+ \frac{1}{2\pi} \int_{o}^{R} (y^{-2} - R^{-2}) \log^+ |f(iy) \, f(-iy)| \, dy + 1/_2 \, |f'(0)|.$$

The right hand side is bounded as a function of R because $\log^+ |f(R e^{i\psi})| \le MR + k$ and $\int_{\mathbb{R}} \frac{\log^+ |f(iy)|}{1 + y^2} dy$ is finite. Thus the set of

$$- \frac{1}{\pi R} \int_{-\pi/2}^{\pi/2} \cos \psi . \log^- |f(R e^{i\psi})| \, d\psi$$

for R not the modulus of a zero of f is bounded above. Hence the set of $\frac{1}{\pi R} \int_{-\pi/2}^{\pi/2} \cos \psi . |\log |f(Re^{i\psi})|| \, d\psi$ is bounded above for all $R \ge 1$. This is seen to be true initially for R not a zero of f, and then this restriction is lifted because near a zero z_n f behaves like a constant times $(z - z_n)^m$ for some positive integer m and $\int_{-1}^{1} |\log|t|| \, dt$ is finite. This completes the proof of Corollary A1.4.

Our next lemma is Poisson's formula for a semicircle, and it will lead to the crucial theorem of Nevanlinna when we let R tend to infinity. Though the proof of the lemma is similar to the proof of Carleman Theorem we give it in detail.

A1.5 LEMMA

Let f be analytic in a neighbourhood of the closed halfplane H^-, and let $z_1, z_2, \ldots$ be the zeros of f in the open right half plane H repeated according to their multiplicities. If R is not the modulus of a zero of f and if $z = x + iy \in H$ with $|z| < R$, then

$$\log|f(z)|$$
$$= \sum_{|z_k|<R} \log\left|\left(\frac{1 - z/z_k}{1 + z/\bar{z}_k} \cdot \frac{R^2 + z z_k}{R^2 - z\bar{z}_k}\right)\right|$$
$$+ \frac{x}{\pi}\int_{-R}^{R} \left\{\frac{1}{v^2 - 2yv + |z|^2} - \frac{R^2}{v^2|z|^2 - 2yvR^2 + R^4}\right\} \log|f(iv)| \, dv$$
$$+ \frac{2Rx}{\pi}\int_{-\pi/2}^{\pi/2} \frac{(R^2 - z^2)\cos\phi}{|Re^{i\phi} - z|^2 \, |Re^{i\phi} + \bar{z}|^2} \log|f(Re^{i\phi})| \, d\phi$$

Proof. Suppose that f has no zero in the half open semidisc $\{z \in H : |z| \le R\}$. Let Γ be the positively oriented closed contour round the semicircle $|z| = R$ in H and along the y-axis between iR and $-iR$ except that the contour detours round each zero of f on the y-axis by a small semicircle in H with centre the zero. As in the proof of Theorem 1.3 we choose and fix a logarithm so that $\log f(z)$ is analytic in a neighbourhood of the closed semidisc $\{z \in H^- : |z| \le R\}$, except for the zeros of f on $i[-R,R]$. We consider the integral

$J = \frac{1}{2\pi i}\int_\Gamma \left\{\frac{1}{w-z} - \frac{1}{w+\bar{z}} - \frac{\bar{z}}{w\bar{z}-R^2} + \frac{z}{wz+R^2}\right\} \log f(w) \, dw$. Grouping the denominator in suitable pairs we see that this is equal to

$$\frac{1}{2\pi i}\int_\Gamma \left\{\frac{2x}{w^2 - 2iyw - |z|^2} - \frac{2R^2 x}{w^2|z|^2 - 2iywR^2 - R^4}\right\} \log f(w) \, dw$$

and also to

$$\frac{1}{2\pi i}\int_\Gamma \left\{\frac{1}{(w-z)(w\bar{z}-R^2)} + \frac{1}{(w+\bar{z})(wz+R^2)}\right\}(|z|^2 - R^2)\log f(w)\,dw.$$

Near one of the zeros α of f on the y-axis $|f(w)|$ behaves like $|w-\alpha|^m$, where m is the order of the zero. Hence the integral along the little semicircle round the zero will behave like a constant times $mt \log t$, where t is the readius of the little semicircle under consideration, and this tends to zero as t tends to zero. We now let the radii of the small semicircles tend to zero and we find the limit of J using the two equivalent forms of J on the y-axis and the semicircle of radius R. The limit is

$$\frac{1}{2\pi}\int_R^{-R}\left\{\frac{2x}{-v^2+2yv-|z|^2} - \frac{2R^2x}{-v^2|z|^2+2yvR^2-R^4}\right\}\log f(iv)\,dv$$

$$+\frac{1}{2\pi i}\int_{-\pi/2}^{\pi/2}\left\{\frac{1}{|Re^{i\phi}+\bar{z}|^2} - \frac{1}{|Re^{i\phi}-z|^2}\;\frac{(|z|^2-R^2)}{re^{i\phi}}\right\}\log f(Re^{i\phi})\,Rie^{i\phi}d\phi.$$

The real part of this is

$$\frac{1}{2\pi}\int_{-R}^{R}\left\{\frac{2x}{v^2-2yv+|z|^2} - \frac{2R^2x}{v^2|z|^2-2yvR^2+R^4}\right\}\log|f(iv)|\,dv$$

$$+\frac{2Rx}{\pi}\int_{-\pi/2}^{\pi/2}\frac{(R^2-|z|^2)\cos\phi}{|Re^{i\phi}-z|^2\,|Re^{i\phi}+\bar{z}|^2}\log|f(Re^{i\phi})|\,d\phi$$

on simplifying the term in brackets in the second integral. The only pole of the integrand in the integral defining J is at $w = z$ caused by the term $(w-z)^{-1}$. From the residue theorem $J = \log f(z)$, and by taking the real part of this the result follows when f has no zeros in $\{z \in H : |z| \le R\}$.

Now suppose that f has n zeros $z_1, \ldots z_n$ in $\{z \in H : |z| < R\}$ with each zero repeated according to its multiplicity. Let $h(z) = \prod_{j=1}^{n} \frac{1 - z/z_j}{1 + z/\bar{z}_j} \cdot \frac{R^2+z_k z}{R^2-\bar{z}_k z}$. Then h is analytic in a neighbourhood of

$\{z \in H^- : |z| \leq R\}$, and by Lemma A1.2 we have $h(z) = 1$ for $z \in i\mathbb{R} \cup \{w \in \mathbb{C} : |w| = R\}$. Let $g(z) = f(z)/h(z)$, then what we have just proved applies to g because g has no zeros in $\{z \in H : |z| < R\}$. The lemma now follows.

A1.6 THEOREM (NEVANLINNA'S FORMULA)

Let f be analytic in a neighbourhood of the closed right half plane H^-, and let f be of exponential type in H with $\gamma = \liminf_{r\to\infty} r^{-1} \log M(r)$, where $M(r) = \sup\{|f(z)| : z \in H, |z| \leq r\}$. If $\int_{\mathbb{R}} \frac{\log^+|f(iw)|dw}{1+w^2}$ is finite, then

$$\log |f(z)|$$

$$= cx + \sum_1^\infty \log \left|\frac{1 - z/z_n}{1 + z/\bar{z}_n}\right| + \frac{x}{\pi}\int_{\mathbb{R}} \frac{\log |f(iw)|}{(w-y)^2 + x^2}\, dw$$

for $z = x + iy \in H$, where $z_1, z_2, \ldots$ are the zeros of f in the open right half plane H repeated according to their multiplicities, and where $c = \lim_{r\to\infty} \frac{2}{\pi r} \int_{-\pi/2}^{\pi/2} \cos\phi . \log |f(re^{i\phi})|\, d\phi$. Further $c \leq 2\gamma$.

Proof. This result is obtained by letting R tend to infinity in the Poisson formula for a semicircle (Lemma A1.5). We shall tackle the various terms on the right hand side of the equation in Lemma A1.5 one-by-one.

The modulus of the difference between the first term

$$\sum_{|z_k|<R} \log \left| \frac{1 - z/z_k}{1 + z/\bar{z}_k} \cdot \frac{R^2 + zz_k}{R^2 - zz_k} \right|$$

of the Poisson formula for a semicircle and $\sum_{|z_k|<R} \log \left|\frac{1 - z/z_k}{1 + z/\bar{z}_k}\right|$ is no greater than

$$\sum_{|z_k|<R} \left|\log \left|\frac{1 - zR^{-2}/z_k}{1 + zR^2/\bar{z}_k}\right|\right| \leq 8|z|R^{-2} \sum_{|z_k|<R} \operatorname{Re}(z_k^{-1})$$

by inequality (2) in the proof of Corollary A1.4 (see p.116) provided R is large enough compared with $|z|$. Since $\sum \mathrm{Re}(z_k^{-1})$ converges, the first term of Lemma A1.5 tends to the $\sum \log \left|\dfrac{1 - z/z_k}{1 + z/\bar{z}_k}\right|$ as R tends to infinity.

The first part of the first integral equals

$$\frac{x}{\pi}\int_{-R}^{R} \frac{1}{w^2-2yw+|z|^2} \log |f(iw)|.\, dw$$

$$= \frac{x}{\pi}\int_{-R}^{R} \frac{1}{x^2+(w-y)^2} \log|f(iw)|dw,$$

which converges to $\dfrac{x}{\pi}\displaystyle\int_{\mathbb{R}} \frac{1}{x^2+(w-y)^2} \log |f(iw)|dw$ as R tends to infinity because $\displaystyle\int_{\mathbb{R}} \frac{|\log|f(iw)||}{1 + w^2}dw$ is finite by Lemma A1.4(iii). Though we do not use it, we note that

$$\frac{x}{\pi}\int_{\mathbb{R}} \frac{1}{x^2+(w-y)^2} \log |f(iw)|dw = (P^x * \log |f(i\cdot)|)(y),$$

where $P^x(y) = \dfrac{x}{(x^2+y^2)}$.

We turn to the second part of the first integral. If R is so large that $|y| < R/2$, and if $G(w) = \dfrac{R^2x}{w^2x^2+R^4/4}$, then

$$\frac{R^2x}{w^2|z|^2-2ywR^2+R^4} = \frac{R^2x}{w^2x^2+(R^2-yw)^2} \le G(w)$$

for $|w| \le R$. Now for $\varepsilon > 0$ given we choose m such that

$$\int_{|w|\ge m} \frac{\log^+|f(iw)|}{1 + w^2} dw \le \varepsilon.$$

Multiplying the integrand by $\dfrac{1+w^2}{1+w^2}$ and splitting the integral at $\pm m$, we obtain

$$\frac{1}{\pi} \int_{-R}^{R} \frac{xR^2}{w^2|z|^2-2ywR^2+R^4} \log |f(iw)| \, dw$$

$$\leq \max \left\{ G(w)(1 + w^2) : |w| \leq m \right\} . \frac{1}{\pi} . \int_{\mathbb{R}} \frac{\log^+|f(iw)|}{1+w^2} \, dw$$

$$+ \frac{\varepsilon}{\pi} \max \left\{ G(w).(1+w^2) : m \leq |w| \leq R \right\}$$

$$\leq \frac{R^2 x \ (m^2+1)}{m^2 x^2+R^4/4} . \frac{1}{\pi} . \int_{\mathbb{R}} \frac{\log^+|f(iw)|}{1+w^2} \, dw + \frac{\varepsilon}{\pi} . 8x$$

since $w \mapsto (1 + w^2) \, G(w)$ is an increasing function of w for $R^4 > 4x^2$ (R is large) and $(1 + R^2)G(R) \leq 8x$. Letting R tend to infinity,

$$\frac{1}{\pi} \int_{-R}^{R} \frac{R^2 x}{w^2|z|^2-2ywR^2+R^4} \log|f(iw)| \, dw$$

tends to zero.

We compare the last integral in A1.5 with $\frac{2x}{\pi r} \int_{-\pi/2}^{\pi/2} \cos\phi \log |f(Re^{i\phi})| \, d\phi$ though we drop the factor $2x$ in the following calculations. Then

$$\left| \frac{1}{\pi R} \int_{-\pi/2}^{\pi/2} \cos \phi . \log|f(Re^{i\phi})| \, d\phi \right.$$

$$\left. - \frac{R}{\pi} \int_{-\pi/2}^{\pi/2} \frac{(R^2 - |z|^2)}{|Re^{i\phi}-z|^2 |Re^{i\phi}+\bar{z}|^2} \cos \phi . \log |f(Re^{i\phi})| \, d\phi \right|$$

$$\leq \frac{1}{\pi R} \int_{-\pi/2}^{\pi/2} \cos \phi . \left|\log |f(Re^{i\phi})|\right| d\phi .$$

$$\max \left\{ \left| 1 - \frac{R^2(R^2 - |z|^2)}{|Re^{i\phi}-z|^2 |Re^{i\phi}+z|^2} \right| : |\phi| \leq \frac{\pi}{2} \right\}$$

for all large R. By Lemma A1.4(iv) the quantity before the maximum is bounded as a function of R. Further the maximum tends to zero as R tend to infinity.

From the above calculations and Lemma A1.5 it follows that the limit giving c exists and that $|\log |f(z)||$ has the required value. Using the inequality $|f(Re^{i\phi})| \leq M(R)$ for $|\phi| \leq \pi/2$ one easily obtains $c \leq 2\gamma$ from the definitions of c and γ. This completes the proof of Theorem A1.6.

The Ahlfors-Heins Theorem is proved by replacing z by $Re^{i\phi}$ in Nevanlinna's formula and letting R tend to infinity. The limits of the terms are easy to calculate except for the term $\sum \log |\cdots|$ arising from the zeros of f essentially via the Blaschke product. The $\sum \log |\cdots|$ term we handle using the inequality in Lemma A1.7, which is due to A.M. Davie.

A1.7 LEMMA

If $0 < \psi < \pi/2$, then there are constants A and B depending on ψ such that

$$-\frac{1}{|u|} \log \left|\frac{u - w}{u + \bar{w}}\right| \leq \frac{\text{Re } w}{|w|^2} (A + B\,|\log|\theta - \phi||)$$

for all $w = \rho e^{i\phi} \in H^-\backslash\{0\}$ and all $u = re^{i\theta} \in H$ with $|\phi| \leq \pi/2$ and $|\theta| \leq \psi$.

Proof. We shall split the pairs (u,w) satisfying the hypotheses into two sets

$$\{(u,w) : |u - w| \geq 1/_2|w| \cos \psi\}, \text{ and}$$

$$\{(u,w) : |u - w| \leq 1/_2|w| \cos \psi\}.$$

We consider these two cases separately.

If $|u - w| \geq 1/_2\, |w| \cos \psi$, then using the inequality $\log (1 + x) \leq x$ for $x > -1$ we obtain

$$\log \left|\frac{u + \bar{w}}{u - w}\right|$$

$$= \frac{1}{2} \log \left|\frac{u + \bar{w}}{u - w}\right|^2$$

$$= \frac{1}{2} \log \left(1 + \frac{4 \operatorname{Re} w \operatorname{Re} u}{|u - w|^2}\right)$$

$$\le 2 \frac{\operatorname{Re} w \operatorname{Re} u}{|u - w|^2} \tag{1}$$

$$\le 8 \sec^2 \psi \frac{|u| \operatorname{Re} w}{|w|^2} .$$

We now consider the more difficult case $|u - w| \le {}^1/_2 |w| \cos$ and we use $u = re^{i\theta}$ and $w = \rho e^{i\phi}$. Squaring the inequality $|u - w| \le {}^1/_2 |w| \cos \psi$ gives

$$r^2 - 2r\rho \cos (\theta - \phi) + \rho^2 \le {}^1/_4 \rho^2 \cos^2 \psi,$$

which leads to

$$\frac{2r \cos \theta \cos \phi}{\rho}$$

$$\ge \frac{r^2}{2} + \frac{2r}{\rho} \sin \theta \sin \phi + 1 - \frac{1}{4} \cos^2 \psi$$

$$\ge \frac{r^2}{\rho^2} - \frac{2r}{\rho} \sin \psi + 1 - \frac{1}{4} \cos^2 \psi,$$

because $|\sin \theta \sin \phi| \le \sin \psi,$

$$\ge \left(\frac{r}{\rho} - \sin \psi^{\,2}\right) + \frac{3}{4} \cos^2 \psi$$

$$\ge \frac{3}{4} \cos^2 \psi.$$

Hence

$$\frac{8}{3} \sec^2 \psi . \ \frac{\operatorname{Re} w}{|w|^2} \ge \frac{1}{|u|} . \tag{2}$$

Using the inequality $t + t^{-1} \ge 2$ for positive $t = r/_\rho$, and the formula for the product of two sines, we have

$$\left|\frac{u+\bar{w}}{u-w}\right| = \frac{r^2 + 2r\rho\cos(\theta+\phi) + \rho^2}{r^2 - 2r\rho\cos(\theta-\phi) + \rho^2}$$

$$= 1 + \frac{4\cos\theta\cos\phi}{r/\rho + \rho/r - \cos(\theta-\phi)}$$

$$\le 1 + \frac{2\cos\theta\cos\phi}{1 - \cos(\theta-\phi)}$$

$$= \frac{1+\cos(\theta+\phi)}{1-\cos(\theta-\phi)} .$$

From the graphs of the functions $1 - \cos\xi$ and $2\xi^2/\pi^2$ for $0 \le \xi \le \pi$, we observe that $1 - \cos\xi \ge 2\xi^2/\pi^2$.

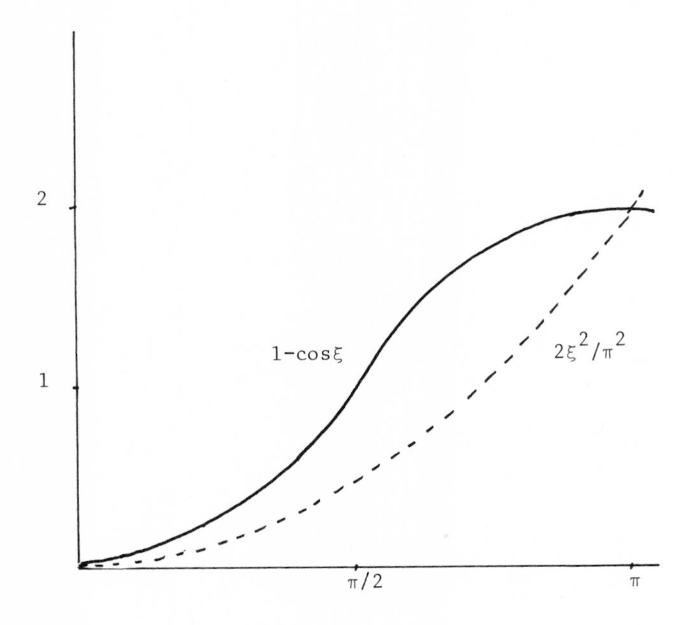

For $0 \le \xi \le \pi/2$ the inequality may be obtained by integrating $\sin\xi \ge 2\xi/\pi$ for $0 \le \xi \le \pi/2$, which gives $1 - \cos\xi \ge \xi^2/\pi \ge 2\xi^2/\pi^2$. The remainder of the range may be checked using the convexity of $1 - \cos\xi$ and the convacity of $2\xi^2/\pi$ on $\pi/2 \le \xi \le \pi$. Thus

$$\log \left|\frac{u + \bar{w}}{u - w}\right|$$

$$\leq \log \left(\frac{2}{1 - \cos(\theta - \phi)}\right) \tag{3}$$

$$\leq \log \left(\frac{\pi^2}{(\theta - \phi)^2}\right)$$

$$\leq 2 \,|\log|\theta - \phi|| + 2 \log \pi.$$

Using (1), (2), and (3) we see that

$$-\frac{1}{|u|} \log \left|\frac{u - w}{u + \bar{w}}\right| \leq \frac{\mathrm{Re}\, w}{|w|^2} (A + B\,|\log|\theta - \psi||),$$

where

$$A = \max\left\{8 \sec^2 \psi, \frac{16}{3} \sec^2 \psi \log \pi\right\} = 8 \sec^2\psi$$

and

$$B = \frac{16}{3} \sec^2 \psi.$$

This proves the lemma.

A1.8 PROOF OF THE AHLFORS-HEINS THEOREM

Let $0 < \psi < \pi/2$. We let $z = re^{i\theta}$ with $|\theta| \leq \psi$ in Nevanlinna's formula, and let r tend to infinity. The term cx in Nevanlinna's formula gives us the limit required in the Ahlfors-Heins' Theorem, and we show that r^{-1} times the other terms tend to zero for almost all θ in $(-\pi/2, \pi/2)$.

We begin with the integral term $J = \frac{x}{\pi} \int_{\mathbb{R}} \frac{\log|f(iw)|}{(w-y)^2+x^2}\, dw$ in Nevanlinna's formula, and split the range of integration into $|w| \leq m$ and $|w| \geq m$ for some large m. In the latter range of integration we use the inequality

$$\frac{\cos\theta}{(w-r\sin\theta)^2 + r^2\cos^2\theta} \le \frac{1}{w^2\cos\theta} \le \frac{1}{w^2\cos\psi} ,$$

which we obtain from $w^2\cos^2\theta \le w^2 - 2rw\sin\theta + r^2$. For large m,

$$\frac{J}{r} \le \int_{|w|\le m} \frac{\log^+|f(iw)|}{w^2 - 2rw\sin\theta + r^2}\, dw + \frac{1}{\cos\psi} \int_{|w|\ge m} \frac{\log^+|f(iw)|}{w^2}\, dw .$$

We now choose, and fix, m so large that $\frac{1}{\cos\psi} \int_{|w|\ge m} \frac{\log^+|f(iw)|}{w^2}\, dw$ is very small. Then for r sufficiently large the integral $\int_{|w|\ge m}$ is also very small. This shows that Jr^{-1} tends to zero uniformly in $|\theta| \le \psi$ as r tends to infinity.

We now tackle the sum of the logarithms in Nevanlinna's formula (that is, the logarithm of the Blaschke product of the zeros of f). Let $\varepsilon > 0$, and let A and B be the constants corresponding to ψ given by Lemma 1.7. Since $\sum \mathrm{Re}(z_n^{-1})$ converges (A1.4), where z_n are the zeros of f in H, we may choose k so that

$$\sum_{n\ge k} \mathrm{Re}(z_n^{-1}) \le \varepsilon^2 .$$

From the equality

$$\left|\frac{1 - re^{i\theta}/z_n}{1 + re^{i\theta}/\bar{z}_n}\right| = \left|\frac{z_n - re^{i\theta}}{\bar{z}_n + re^{i\theta}}\right|$$

and Lemma 1.7 with $u = re^{i\theta}$ and $w = z_n = |z_n|\, e^{i\phi_n}$, we obtain

$$\begin{aligned} F(re^{i\theta}) &\equiv -\frac{1}{r} \sum_{n\ge k} \log \left|\frac{1 - re^{i\theta}/z_n}{1 + re^{i\theta}/\bar{z}_n}\right| \\ &\le \sum_{n\ge k} \frac{\mathrm{Re}\, z_n}{|z_n|^2} \left\{A + B \log |\theta - \phi_n|\right\} \\ &\equiv G(\theta) \end{aligned}$$

for all $r > 0$ and $\theta \in [-\psi, \psi]$. Let U be the set of all $\theta \in [-\psi, \psi]$

such that

$$\limsup_{r\to\infty} -\frac{1}{r}\sum_{n=1}^{\infty} \log \left|\frac{1 - re^{i\theta}/z_n}{1 + re^{i\theta}/\bar{z}_n}\right| \geq 2\varepsilon.$$

Since

$$\lim_{r\to\infty} \log \left|\frac{1 - re^{i\theta}/z_n}{1 + re^{i\theta}/\bar{z}_n}\right| = 0$$

for $1 \leq n \leq k$, we have $\theta \in U$ if and only if $\limsup_{r\to\infty} F(re^{i\theta}) \geq 2\varepsilon$.

Thus

$$U \subseteq \{\theta \in [-\psi,\psi] : F(re^{i\theta}) \geq \varepsilon \text{ for some } r\}$$
$$\subseteq \{\theta \in [-\psi,\psi] : G(\theta) \geq \varepsilon\}.$$

If μ is Lebesgue measure on $\mathbb{R}$, then

$$\varepsilon.\mu(U) \leq \varepsilon.\mu\{\theta \in [-\psi,\psi] : G(\theta) \geq \varepsilon\}$$
$$\leq \int_{-\psi}^{\psi} G(\theta)d\theta$$
$$= \sum_{n\geq k} \frac{\text{Re}z}{|z_n|^2} \int_{-\psi}^{\psi} (A + B\,|\log|\theta - \phi_n||d\theta$$
$$\leq \varepsilon^2\,(\pi A + 2B\int_0^{\pi/2} |\log t|dt)$$

since $|\theta - \phi_n| \leq \psi \leq \pi/2$ for all n. Therefore $\mu(U) \leq C\varepsilon$, where C is a constant depending on A and B but independent of ε. Thus

$$\limsup_{r\to\infty} -\frac{1}{r}\sum_{1}^{\infty} \log \left|\frac{1 - re^{i\theta}/z_n}{1 + re^{i\theta}/\bar{z}_n}\right| \leq 0$$

for almost all $\theta \in [-\psi,\psi]$. Because $re^{i\theta}$ and z_n are in H, we have $\left|\frac{1 - re^{i\theta}/z_n}{1 + re^{i\theta}/\bar{z}_n}\right| \leq 1$, and hence $-\frac{1}{r}\sum_{1}^{\infty} \log \left|\frac{1 - re^{i\theta}/z_n}{1 + re^{i\theta}/\bar{z}_n}\right| \geq 0$ for all $\theta \in (-\pi/2,\pi/2)$. Thus the lim sup above is equal to the limit almost everywhere and is zero almost everywhere. The proof of the Ahlfors-Heins' Theorem is complete.

The following corollary of Nevanlinna's formula will be used in Appendix 2 when we investigate closed ideals in $L^1(\mathbb{R}^+,\omega)$. This corollary is a weaker version of a theorem of Krein that an entire function which is the quotient of bounded analytic functions in the upper and lower half planes is of exponential type.

A1.9 COROLLARY

Let f be an entire function such that f is bounded in the left half plane $-H$. If there are functions g and k analytic in a neighbourhood of the closed right half plane H^- with $fg = k$ and g and k bounded in H^-, then f is of exponential type in $\mathbb{C}$.

Proof. Since f is bounded in the left half plane, we may assume that $|f(iw)| \leq 1$ for all $w \in \mathbb{R}$. We apply Nevanlinna's formula to g and k, where $z_1, z_2, \ldots$ and $w_1, w_2, \ldots$ are the zeros of g and k in the open right half plane H with each zero repeated according to its multiplicity. Since $f(z) = k(z)/g(z)$ for z not a zero of g and since f is an entire function, the zeros of g are zeros of k of smaller multiplicity. Hence

$$\sum_1^\infty \left\{\log\left|\frac{1 - z/w_n}{1 + z/\bar{w}_n}\right| - \log\left|\frac{1 - z/z_n}{1 + z/\bar{z}_n}\right|\right\} = \sum \log\left|\frac{1 - z/w_{n_j}}{1 + z/\bar{w}_{n_j}}\right| ,$$

where the second summation is over those zeros of k not cancelled by zeros of g, that is, over the zeros of f in H. We now apply Nevanlinna's formula to g and k obtaining, for some constant C and all $z = x+iy \in H$,

$$\begin{aligned}
&\log|f(z)| \\
&= \log|k(z)| - \log|g(z)| \\
&= Cx + \sum_1^\infty \left\{\log\left|\frac{1 - z/w_n}{1 + z/\bar{w}_n}\right| - \log\left|\frac{1 - z/z_n}{1 + z/\bar{z}_n}\right|\right\} \\
&\qquad + \frac{x}{\pi}\int_{\mathbb{R}} \frac{\log|k(iw)| - \log|g(iw)|}{(w-y)^2 + x^2}\, dw \\
&\leq C\,x + \sum_1^\infty \log\left|\frac{1 - z/w_{n_j}}{1 + z/\bar{w}_{n_j}}\right| + \frac{x}{\pi}\int_{\mathbb{R}} \frac{\log|f(iw)|}{(w-y)^2 + x^2}\, dw \\
&\leq Cx,
\end{aligned}$$

because $|f(iw)| \le 1$ for all $w \in \mathbb{R}$ and $\left|\frac{1 - z/w_{n_j}}{1 + z/\bar{w}_{n_j}}\right| \le 1$ for all j.

Hence f is of exponential type in the right half plane, and so in $\mathbb{C}$.

A1.10 NOTES AND REMARKS

The Ahlfors-Heins Theorem is in Ahlfors and Heins [1949], and Boas [1954] has a section devoted to it (Section 7.2) - see also Hayman [1956]. However there are two minor points in Boas's account that make matters difficult for the innocent reader. Boas omits the factor $(R^2 - z_n z)/(R^2 + \bar{z}_n z)$ in many of the calculations - the factor in his case is $(R^2 - z_n z)/(R^2 - \bar{z}_n z)$ because he is working in the upper half plane (see the list of errata in Boas [1974]). Also he works sometimes in the upper half plane and sometimes in the lower half plane, and though there is nothing wrong it adds further confusion. The account here is influenced by Boas [1954] (Section 7.2). Lemma A1.7 and this variation of the proof of the Ahlfors-Heins Theorem are due to A.M. Davie, and I am grateful for his permission to use these ideas here. Krein's Theorem is in Krein [1947], and I do not know of a simple published proof.

APPENDIX 2: ALLAN'S THEOREM - CLOSED IDEALS IN $L^1(\mathbb{R}^+,\omega)$

A major open problem in radical Banach algebra theory is, are all the proper closed ideals in $L^1(\mathbb{R}^+,\omega)$ equal to the standard ideals $I_\beta = \{g \in L^1(\mathbb{R}^+,\omega) : g = 0 \text{ a.e. on } [0,\beta]\}$ for suitable radical weights ω? We shall prove a theorem of Allan [1979] that certain closed ideals in $L^1(\mathbb{R}^+,\omega)$ are of the standard form. We use this theorem in Chapter 4. We firstly recall some notation and ideas from 2.12. A radical weight ω on $\mathbb{R}^+$ is a continuous positive valued function $\omega : \mathbb{R}^+ = [0,\infty) \to (0,\infty)$ satisfying $\omega(s + t) \le \omega(s)\,\omega(t)$ for all $s, t > 0$ and $\omega(r)^{1/r} \to 0$ as $r \to \infty$. The Banach space $L^1(\mathbb{R}^+,\omega) = \{g : g$ is measurable, $\int_0^\infty |g(t)|\,\omega(t)\,dt = \|g\| < \infty\}$ becomes a radical Banach algebra under the convolution product $(f * g)(t) = \int_0^t f(t - w)\,g(w)\,dw$ for a.e. $t \in \mathbb{R}^+$.

A2.1 THEOREM

Let ω be a radical weight on $\mathbb{R}^+$, and let J be a closed ideal in $L^1(\mathbb{R}^+,\omega)$. If there is a non-zero $f \in J$ and a $k > 0$ such that $\int_0^\infty |f(t)|\,e^{-kt}\,dt$ is finite, then there is a $\gamma \ge 0$ such that

$$J = \{g \in L^1(\mathbb{R}^+,\omega) : g = 0 \text{ a.e. on } [0,\gamma]\}.$$

Proof. If $0 \le \beta$, let

$$I(\beta) = \{g \in L^1(\mathbb{R}^+,\omega) ; g = 0 \text{ a.e. on } [0,\beta]\}.$$

A direct calculation shows that each $I(\beta)$ is a closed ideal in $L^1(\mathbb{R}^+,\omega)$, and clearly $I(0) = L^1(\mathbb{R}^+,\omega)$. Let $\gamma = \sup\{\beta : I(\beta) \supseteq J\}$. If $\gamma > 0$, $I(\gamma) = \cap\{I(\beta) : \beta < \gamma\}$ so that $I(\gamma) \supseteq J$. If $\gamma = 0$, clearly $I(\gamma) \supseteq J$. We shall prove the reverse inclusion.

Let $L^\infty(\mathbb{R}^+,\omega^{-1}) = \{h : h \text{ measurable}, \|h/\omega\|_\infty = \|h\| < \infty\}$. Then $L^\infty(\mathbb{R}^+,\omega^{-1})$ is a Banach space, and there is a natural duality between $L^1(\mathbb{R}^+,\omega)$ and $L^\infty(\mathbb{R}^+,\omega^{-1})$ given by $<g,h> = \int_0^\infty h(w)\, g(w)\, dw$ for $g \in L^1(\mathbb{R}^+,\omega)$ and $h \in L^\infty(\mathbb{R}^+,\omega^{-1})$ identifying $L^\infty(\mathbb{R}^+,\omega^{-1})$ with the dual space $L^1(\mathbb{R}^+,\omega)^*$ of $L^1(\mathbb{R}^+,\omega)$. We define the function α on the space of measurable functions on $\mathbb{R}^+$ by $\alpha(g)$ is the infimum of the support of g.

Let $f_1 \in J$ and let $K \in \mathbb{R}$ such that $\int_0^\infty |f_1(t)|\, e^{(-K+1)t}\, dt$ is finite. Let $g \in L^\infty(\mathbb{R}^+,\omega^{-1})$ with $<g, f_1 * L^1(\mathbb{R}^+,\omega)> = \{0\}$. We let $F(t) = e^{-Kt} f_1(t)$ and $G(t) = e^{Kt} g(t+1)$ for all $t \in \mathbb{R}^+$. We identify $L^1(\mathbb{R}^+)$ with the Banach subalgebra of $L^1(\mathbb{R})$ of functions that vanish on $(-\infty,0)$. We define $\tilde{}$ on $L^1(\mathbb{R})$ by $\tilde{h}(w) = h(-w)$ for all $w \in \mathbb{R}$. Since $\omega(t)^{1/t} \to 0$ as $t \to \infty$ and $|g(t)| \le \|g\|\, \omega(t)$ for almost all $t > 0$, we have $\tilde{G} \in L^1(\mathbb{R}) \cap L^2(\mathbb{R})$ and the Fourier transform $\tilde{G}^\wedge(\lambda/2\pi) = \int_{-\infty}^0 \tilde{G}(t)\, e^{-i\lambda}$ is an entire function that is bounded in the closed upper half plane $\Pi = \{z \in \mathbb{C} : \operatorname{Im} z \ge 0\}$. Further the Fourier transform $F^\wedge$ is analytic in a neighbourhood of $-\Pi$ and $F^\wedge$ is bounded on $-\Pi$ because F is zero on $(-\infty,0)$ and $t \mapsto e^{(-K+1)t} f(t)$ is integrable on $[0,\infty)$.

For each $x > 0$ the function $t \mapsto f_1(t-x) : [0,\infty) \to \mathbb{C}$ is in $(f_1 * L^1(\mathbb{R}^+,\omega))^-$ because $t \mapsto f_1(t-x)$ is the norm limit of $f_1 * e_n$, where e_n is $2n$ times the characteristic function of $[x - 1/n, x + 1/n]$. Thus $\int_0^\infty f_1(t)\, g(t+x)dt = \int_0^\infty f_1(t-x)\, g(t)\, dt = 0$ for all $x > 0$ so th

$$F * \tilde{G}(-w) = \int_0^\infty F(t)\, G(t+w)\, dt$$

$$= \int_0^\infty e^{-Kt} f(t)\, e^{Kt}.e^{Kw} g(t+w+1)\, dt = 0$$

for all $w \ge -1$.

Let $L = F * \tilde{G}$ in $L^1(\mathbb{R})$. Then L is zero on $(-\infty,1]$. Hence the Fourier transform $L^\wedge$ is analytic in a neighbourhood $\{z \in \mathbb{C} : \operatorname{Im} z < 1\}$ of the closed lower half plane $-\Pi$, and is bounded in $-\Pi$. The entire function $\tilde{G}^\wedge$ satisfies the hypotheses of Corollary A1.9 if we rotate the complex pla through $\pi/2$ and identify the closed lower half plane $-\Pi$ with the closed right half plane H^-. Thus $\tilde{G}^\wedge$ is of exponential type. Using $\tilde{G}$, and so $\tilde{G}^\wedge$, is in $L^2(\mathbb{R})$ we may apply the Paley-Wiener Theorem (see Rudin [1966]) and deduce that $\tilde{G}$, and so g, has compact support. Let β be the supremum of the support of g. From the definition of $\tilde{G}(t) = g(-t+1)e^{-Kt}$ it follows that $\alpha(G) = -\beta + 1$. The Titchmarsh Convolution Theorem (see Mikusinski [1959]) states that $\alpha(F * \tilde{G}) = \alpha(F) + \alpha(\tilde{G})$. Since $F * \tilde{G}$ is zero on $(-\infty, 1]$, $\alpha(F * \tilde{G}) \geq 1$. Thus $1 \leq \alpha(f_1) - \beta + 1$, and therefore the support of g is contained in the interval $[0, \alpha(f_1)]$.

Since g is an arbitrary function in $L^\infty(\mathbb{R}^+, \omega^{-1})$ orthogonal to $(f_1 * L^1(\mathbb{R}^+,\omega))^-$ it follows that $(f_1 * L^1(\mathbb{R}^+,\omega))^- = I(\alpha(f_1))$ by the isomorphism between $L^1(\mathbb{R}^+,\omega)^*$ and $L^\infty(\mathbb{R}^+,\omega^{-1})$.

Using f in place of f_1 we obtain $J \supseteq I(\alpha(f))$ because $f \in J$. If $\alpha(f) = \gamma$, we are done so we suppose that $\gamma < \alpha(f)$. Since $\gamma = \inf\{\alpha(h) : h \in J\}$, for $\varepsilon > 0$ there is an $h \in J$ such that $\alpha(h) \leq \gamma + \varepsilon$. Now the restriction of h to $[\alpha(f), \infty)$ is in $I(\alpha(f))$, so in J, and thus the restriction of h to $[\gamma, \alpha(f)]$ is in $J \cap L^1(\mathbb{R}^+)$. Therefore $J \supseteq (h * L^1(\mathbb{R}^+,\omega))^- = I(\alpha(h)) \supseteq I(\gamma + \varepsilon)$. Since $\bigcup_{\varepsilon>0} I(\gamma+\varepsilon)$ is dense in $I(\gamma)$, we have $J \supseteq I(\gamma)$. The proof is complete.

A2.2 NOTES AND REMARKS

This is one of the two results in Allan [1979] showing that if there is a nice f in a closed ideal J of $L^1(\mathbb{R}^+,\omega)$, then J is standard. The other condition is a local one near zero. See Dales [1978] for a general discussion. These results have been strengthened by Domar [1981].

APPENDIX 3: QUASICENTRAL BOUNDED APPROXIMATE IDENTITIES

In Chapters 3 and 4 we used bounded approximate identities, which had nice properties with respect to derivations, multipliers, or automorphisms, to obtain corresponding properties for an analytic semigroup $t \mapsto a^t$ near 0 in $\mathbb{R}^+$. The problem is when do these nice bounded approximate identities exist? In this appendix we show that in an Arens regular Banach algebra with a bounded approximate identity, there is a new bounded approximate identity that behaves well with respect to derivations, multipliers, and automorphisms. The idea is to lift the problem from the algebra A to the second dual A^{**} with the Arens product and exploit the identity in A^{**} given by the hypotheses.

A3.1 <u>DEFINITION</u>

Let A be a Banach algebra, and define the Arens product $F.G$ on A^{**} as follows

$$\langle f.x, y\rangle = \langle f, xy\rangle$$

$$\langle F.f, x\rangle = \langle F, f.x\rangle$$

$$\langle F.G, f\rangle = \langle F, G.f\rangle$$

for all $x,y \in A$, $f \in A^*$, and $F,G \in A^{**}$, where $\langle \cdot,\cdot \rangle$ is the pairing between dual Banach spaces.

The reversed product $\circ$ on the Banach algebra A is defined by $x \circ y = yx$ for all $x,y \in A$. The Banach algebra A is said to be Arens regular if the reverse of the Arens product in A^{**} is the Arens product of the reversed product in A. With a little calculation we could write down the definition of Arens regularity symbolically and relate it to the relationship between A and A^*, but it is the idea we require. Let $\wedge$ be the natural embedding from A into A^{**} given by $\langle x^\wedge, f\rangle = \langle f, x\rangle$

for all $x \in A$ and $F \in A^*$.

A3.2 LEMMA

Let A be a Banach algebra with a bounded approximate identity Λ. If A is Arens regular, then A^{**} has an identity E, and E is in the $\sigma(A^{**},A^*)$-closure of $\Lambda^\wedge$.

Proof. Since the unit sphere of A^{**} is compact in the $\sigma(A^{**},A^*)$ - topology, there is a net $\{e_\gamma : \gamma \in \Gamma\}$ in Λ such that $\{e_\gamma : \gamma \in \Gamma\}$ is a bounded approximate identity for A, and $e_\gamma^\wedge$ tends to E in the $\sigma(A^{**},A^*)$ - topology. If $f \in A^*$, then $\langle E.f,x\rangle = \langle E,f.x\rangle = \lim \langle e_\gamma, f.x\rangle$ $= \lim \langle f.x,e_\gamma\rangle = \lim \langle f,x.e_\gamma\rangle = \langle f,x\rangle$ for all $x \in A$ so that $E.f = f$. Hence $F.E = F$ for all $F \in A^{**}$, and E is a right identity for A^{**}. Since $\{e_\gamma : \gamma \in \Gamma\}$ is also a left bounded approximate identity for A, we have $F \square E = F$ for all $F \in A$ where $\square$ is the Arens product in A^{**} arising from the reversed product in A. When A is Arens regular $F \square E = E.F$, so E is an identity for A^{**}. This proves the lemma.

Recall that a derivation D on an algebra A is a linear operator on A satisfying $D(ab) = aD(b) + D(a)b$ for all $a,b \in A$.

A3.3 THEOREM

Let A be a Banach algebra with a bounded approximate identity Λ, let $\mathrm{Aut}(A)$ denote the group of continuous automorphisms on A, and let $\mathrm{Der}(A)$ denote the Banach space of continuous derivations on A. If A is Arens regular, then for each $\varepsilon > 0$ and all finite subsets $F \subseteq A$, $G \subseteq \mathrm{Aut}(A)$, and $\mathcal{D} \subseteq \mathrm{Der}(A)$, there is an $e \in \mathrm{co}(\Lambda)$, the convex hull of Λ, such that $\|e.x - x\| + \|x.e - x\| < \varepsilon$, $\|\alpha(e) - e\| < \varepsilon$, and $\|D(e)\| < \varepsilon$ for all $x \in F$, $\alpha \in G$, and $D \in \mathcal{D}$.

Proof. Let $\{e_\gamma : \gamma \in \Gamma\}$ be the net in Λ given by Lemma A3.2 such that $e_\gamma^\wedge$ tends to E in the $\sigma(A^{**},A^*)$-topology. Choose $\beta \in \Gamma$ such that $\|e_\gamma.x - x\| + \|x.e_\gamma - x\| < \varepsilon$ for all $x \in F$ and all $\gamma \in \Gamma$ with $\gamma > \beta$. If $G = \{\alpha_1,\dots,\alpha_m\}$ and $\mathcal{D} = \{D_1,\dots,D_n\}$, let W be the convex hull of the set

$$\{(1 - \alpha_1)e_\gamma,\cdots,(1 - \alpha_m)e_\gamma,D_1(e_\gamma),\cdots,D_n(e_\gamma)) : \gamma \in \Gamma,\ \gamma > \beta\}$$

in the Banach space A^{m+n} with norm $\|(x_j)\| = \max\{\|x_j\| : 1 \le j \le m + n\}$.

The norm closure of W and the weak closure of W coincide by the Hahn-Banach Theorem, and we shall show that (0) is in this closure.

A straightforward calculation using the definition of the Arens product on A^{**} and of the second dual of an operator shows that α^{**} is an automorphism on A^{**} for $\alpha \in \mathrm{Aut}(A)$, and D^{**} is a derivation on A^{**} for $D \in \mathrm{Der}(A)$. Since E is the identity in A^{**}, we have $(\alpha^{**} - 1)(E) = 0 = D^{**}(E)$. For each $f \in A^*$ we have

$$
\begin{aligned}
0 &= \langle(\alpha^{**} - 1)E, f\rangle \\
&= \langle E, (\alpha^* - 1)f\rangle \\
&= \lim \langle(\alpha^* - 1)f, e_\gamma\rangle \\
&= \lim \langle f, (\alpha - 1)e_\gamma\rangle ,
\end{aligned}
$$

and

$$0 = \langle D^{**}(E), f\rangle = \lim \langle f, D(e_\gamma)\rangle .$$

Hence (0) is in the weak closure of W in A^{m+n}. Thus there is a convex combination e from $\{e_\gamma : \gamma \in \Gamma, \gamma > \beta\}$ such that $\|(\alpha - 1)e\| < \varepsilon$ and $\|D(e)\| < \varepsilon$ for all $\alpha \in \mathcal{G}$ and $D \in \mathcal{D}$. The proof is complete.

If the hypotheses of the above theorem are satisfied, then A has a quasicentral bounded approximate identity for each enveloping Banach algebra containing A as a two-sided ideal (see 4.8, proof of property 17) Since C*-algebras are Arens regular (see Civin and Yood [1961]) this gives another way of writing the proof that a C*-algebra has a quasicentral bounded approximate identity (see Arveson [1977], Akemann and Pedersen [1977], and Sinclair [1979a]). Suppose that a Banach algebra A has a bounded approximate identity. Then A has a quasicentral bounded approximate identity if

(i) $A^* = A.A^* + A^*.A$ (see Sinclair [1979a]), or

(ii) A has a central bounded approximate identity (trivial), or

(iii) A is Arens regular

If $B = (A + C)^-$, where C is a commutative Banach algebra generated as a Banach algebra by the set $\mathcal{U} = \{u \in C : \|u\| = \|u^{-1}\| = 1\}$, then A has a quasicentral bounded approximate identity. The proof is similar to A3.3 except that the element E comes from the Markov-Kakutani fixed point theorem not from A3.2, and $E \in A^{**}$ commutes with all $u \in \mathcal{U}$ rather than being an identity for A^{**}.

A3.4 PROBLEM

Let G be a locally compact group. When does $L^1(G)$ have a quasicentral bounded approximate identity?

A3.5 NOTES AND REMARKS

Lemma A3.2 is in Civin and Yood [1961]. Quasicentral bounded approximate identities play a fundamental role in C*-algebra theory. See Arveson [1977], Akemann and Pedersen [1977], and Elliott [1977].

REFERENCES

J.F. Aarnes and R.V. Kadison. Pure states and approximate identities, Proc. Amer. Math. Soc. 21 (1969) 749-752.

L. Ahlfors and M. Heins. Questions of regularity connected with the Phragmén-Lindelöf principle, Ann. of Math. (2) 50 (1949) 341-346.

C.A. Akemann and G.K. Pedersen. Ideal perturbations of elements in C*-algebras, Math. Scand. 41 (1977) 117-139.

G.R. Allan. Ideals of rapidly growing functions, Proceedings International Symposium on functional analysis and its applications, Ibadan, Nigeria (1977) 85-109.

G.R. Allan and A.M. Sinclair. Power factorization in Banach algebras with a bounded approximate identity, Studia Math. 56 (1976) 31-38.

W. Arveson. Notes on extensions of C*-algebras, Duke Math. J. 44 (1977) 329-355.

W.G. Bade and H.G. Dales. Norms and ideals in radical convolution algebras, J.Functional Analysis 41 (1980) 77-109.

C. Berg and G. Forst. Potential theory on locally compact abelian groups, Springer-Verlag, Berlin, 1975.

R.P. Boas. Entire functions, Academic Press, New York, 1954.

S. Bochner. Harmonic analysis and the theory of probability, University of California Press, California, 1955.

F.F. Bonsall and J. Duncan. Complete normed algebras, Springer-Verlag, Berli 1973.

P.L. Butzer and H. Berens. Semi-groups of operators and approximation, Springer-Verlag, Berlin, 1967.

M.D. Choi and E.G. Effros. Nuclear C*-algebras and the approximation property, Amer. J. Math. 100 (1978) 61-79.

P.Civin and B. Yood. The second conjugate space of a Banach algebra as an algebra, Pacific J. Math. 11 (1961) 847-870.

P.J. Cohen. Factorization in group algebras, Duke J. Math. 26 (1959) 199-205.

H.S. Collins and W.H. Summers. Some applications of Hewitt's factorization theorem, Proc. Amer. Math. Soc. 21 (1969) 727-733.

H.G. Dales. Automatic continuity : a survey, Bull. London Math. Soc. 10 (1978) 129-183.

J. Dixmier. Les opérateurs permutables á l'operateur integral, Portugal. Math. 8 (1949) 73-84.

J. Dixmier. Opérateurs de rang fini dans les représentations unitairés, Publ. Math. I.H.E.S. 6 (1960) 13-25.

J. Dixmier. Les C*-algebrès et leurs representations, Gauthier-Villars, Paris, 1969.

P.G. Dixon. Approximate identities in normed algebras, Proc. London Math. Soc. (3) 26 (1973) 485-496.

P.G. Dixon. Approximate identities in normed algebras II, J. London Math. Soc. 17 (1978) 141-151.

P.G. Dixon. Spectra of approximate identities in Banach algebras, Math. Proc. Camb. Philos. Soc. 86 (1979) 271-278.

R.S. Doran and J. Wichman. Approximate identities and factorization in Banach modules, Lecture Notes in Math. 768, Springer-Verlag, Berlin, 1979.

N. Dunford and J.T. Schwartz. Linear Operators Part I, J.Wiley, New York, 1958.

G.A. Elliott. Some C*-algebras with outer derivations III, Ann. of Math. 106 (1977) 121-143.

A. Erdelyi, et al. Higher transcendental functions, Vol. 1, Bateman Manuscript Project, McGraw-Hill, New York, 1953.

A. Erdelyi, et al. Tables of integral transforms, Vol. 1, Bateman Manuscript Project, McGraw-Hill, New York, 1954.

J. Esterle. Injection de semigroups divisibles dans des algebres de convolution et construction d'homomorphismes discontinus de C(K), Proc. London Math. Soc. (3) 36 (1978) 46-58.

J. Esterle. Theorems of Gelfand-Mazur type and continuity of epimorphisms from C(X), J. Functional Analysis 36 (1980) 273-286 [1980a].

J. Esterle. Infinitely divisible elements and universal properties for commutative radical Banach algebras which possess elements of finite closed descent, [1980b] to appear.

J. Esterle. Nonseparability and disconnectedness of the group of invertible elements for the multiplier algebra of some commutative Banach algebras with bounded approximate identities, [1980c] to appear.

J. Esterle. Irregularity of the rate of decrease of sequences of powers in the Volterra algebra, [1980d] to appear.

J. Esterle. Rates of decrease of sequences of powers in commutative radica Banach algebras, [1980e] Pacific J. Math. 94 (1981) 61-82.

J. Esterle. A complex variable proof of the Wiener Tauberian Theorem, Ann. Institute Fourier, Grenoble 30 (1980) 91-96.

J. Esterle. Elements for a classification of commutative radical Banach algebras [1980g]

D.E. Evans and J.T. Lewis. Some semigroups of completely positive maps on the CAR algebra, J. Functional Analysis 26 (1977) 369-377.

I. Gelfand, R. Raikov, and G. Shilov. Commutative normed rings, Chelsea, New York, 1964.

F. Ghahramani. Homomorphisms and derivations on weighted convolution algebras, J. London Math. Soc. 21 (1980) 149-161.

S. Grabiner, Derivations and automorphisms of Banach algebras of power series, Memoir Amer. Math. Soc. 146, Amer. Math. Soc., Providence, R.I., 1974.

S. Grabiner, Unbounded multipliers of convergent series in Banach spaces, [1980] to appear.

G.H. Hardy and J.E. Littlewood. Some properties of fractional integrals II Math. Z. 4 (1932) 403-439.

W.K. Hayman, Questions of regularity connected with the Phragmén-Lindelöf principle, J. Math. Pure et Appl. (9) 35 (1956) 115-126.

D.J. Hebert and H.E. Lacey. On supports of regular Borel measures, Pacific Math. 27 (1968) 101-118.

E. Hewitt and K.A. Ross. Abstract harmonic analysis I, Springer-Verlag, Berlin, 1963.

E. Hewitt and K.A. Ross. Abstract harmonic analysis II, Springer-Verlag, Berlin, 1970.

E. Hille and R.S. Phillips. Functional analysis and semigroups, Revised Ed American Math. Soc., Providence, R.I., 1974.

J. Hoffman-Jørgensen. The theory of analytic spaces, Various publications series 10, Aarhus University, Aarhus, Denmark, 1970.

G. Hunt. Semigroups of measures on Lie groups, Trans. Amer. Math. Soc. 81 (1956) 264-293.

A. Hulanicki. On the spectrum of convolution operators on groups with polynomial growth, Inventiones Math. 17 (1972) 135-142.

A. Hulanicki. Subalgebra of $L_1(G)$ associated with Laplacian on a Lie group, Colloquium Math. 31 (1974) 259-287.

A. Hulanicki and T. Pytlik. On commutative approximate identities and cyclic vectors of induced representations, Studia Math. 48 (1973) 189-199.

B.E. Johnson. An introduction to the theory of centralizers, Proc.London Math. Soc. 14 (1964) 299-320.

B.E. Johnson. Continuity of centralizers on Banach algebras, J. London Math. Soc. 41 (1966) 639-640.

B.E. Johnson. Cohomology in Banach algebras, Mem. Amer. Math. Soc. 127, Amer. Math. Soc., Providence, R.I., 1972.

M.G. Krein. A contribution to the theory of entire functions of exponential type, Izvestiya Akad. Nauk S.S.S.R. 11 (1947) 309-326 (in Russian).

E.C. Lance. On nuclear C*-algebras, J. Functional Analysis 12 (1973) 157-176.

H. Leptin. On group algebras of nilpotent groups, Studia Math. 47 (1973) 37-49.

H. Leptin. Ideal theory in group algebras of locally compact groups, Inventiones Math. 31 (1976) 259-278.

N. Levinson. Gap and density theorems, American Math. Soc., Providence, R.I., 1939.

R.J. Loy, Notes on analytic spaces, [1978] preprint.

J. Mikusiński. Operational calculus, Pergamon Press, London, 1959.

F.W.J. Olver. Asymptotics and special functions, Academic Press, New York, 1974.

T.W. Palmer. Classes of non-abelian, noncompact, locally compact groups, Rocky Mt. J. Math. 8 (1978) 683-741.

G.K. Pedersen. C*-algebras and their automorphism groups, Academic Press, London, 1979.

T. Pytlik. On commutative approximate identities, Studia Math. 53 (1975) 265-267.

H. Radjavi and P. Rosenthal. Invariant subspaces, Springer-Verlag, Berlin, 1973.

M. Reed and B. Simon. Methods of modern mathematical physics, Vol.2, Fourier analysis, self-adjointness, Academic Press, New York, 1975.

C.E. Rickart. General theory of Banach algebras, New York, Van Nostrand, 1960.

W. Rudin. Real and complex analysis, McGraw-Hill, New York, 1966.

A.M. Sinclair. Bounded approximate identities, factorization, and a convolution algebra, J. Functional Analysis 29 (1978) 308-318.

A.M. Sinclair. Cohen's factorization method using an algebra of analytic functions, Proc. London Math. Soc. 39 (1979) 451-468 [1979a].

A.M. Sinclair. Cohen elements in Banach algebras. Proc. Royal Society of Edinburgh 84 A (1979) 55-70 [1979b].

E.M. Stein. Topics in harmonic analysis related to the Littlewood-Paley theory, Princeton Univ. Press, Princeton, N.J., 1970.

E.M. Stein Singular integrals and differentiability properties of functic Princeton Univ. Press, Princeton, N.J., 1970.

E.M. Stein and G. Weiss. Introduction to Fourier analysis on Euclidean spaces, Princeton Univ. Press, Princeton, N.J., 1971.

F. Stinespring. Positive functions on C*-algebras, Proc. American Math. Sc 6 (1959) 211-216.

E.C. Titchmarsh. The theory of functions, 2nd ed., Oxford Univ. Press, Oxford, 1939.

N.T. Varopoulos. Continuité des formes linéaires positives sur une algèbre de Banach avec involution, C.R. Acad. Sci. Paris, Serie A, 258 (1964) 1121-1124.

D.V. Widder. The heat equation, Academic Press, New York, 1975.

REFERENCES ADDED IN PROOF

H.G. Dales and W.K. Hayman. Esterle's proof of the Tauberian theorem for Beurling algebras, Ann. Inst: Fourier (to appear) [1981].

Y. Domar. Extensions of the Titchmarsh convolution theorem with applicatio in the theory of invariant subspaces, (to appear) [1981].

U. Haagerup. The Grothendieck inequality for bilinear forms on C^*-algebras (to appear) [1981].

G. Pisier. Holomorphic semigroups and the geometry of Banach spaces, Ann. Math. (to appear) [1981].

INDEX

NOTE. The terms "analytic semigroup", "Banach algebra", "bounded approximate identity", and "semigroup" occur so frequently they are not covered in the index except for the first occurrence.

The most important page number is underlined.

LONDON MATHEMATICAL SOCIETY
LECTURE NOTE SERIES

Edited by PROFESSOR I. M. JAMES
Mathematical Institute, 24-29 St Giles, Oxford

with the assistance of
F. E. Browder *(Chicago)*
M. W. Hirsch *(Berkeley)*
N. Katz *(Princeton)*
G.-C. Rota *(M.I.T.)*
D. S. Scott *(Carnegie-Mellon)*
S. T. Yau *(Institute of Advanced Study, Princeton)*

Continuous Semigroups in Banach Algebras

ALLAN M. SINCLAIR
Reader in Mathematics, University of Edinburgh

In these notes the abstract theory of analytic one-parameter semigroups in Banach algebras is discussed, with the Gaussian, Poisson and fractional integral semigroups in convolution Banach algebras serving as motivating examples. Such semigroups are constructed in a Banach algebra with a bounded approximate identity. Growth restrictions on the semigroup are linked to the structure of the underlying Banach algebra. The Hille-Yosida Theorem and a result of J. Esterle's on the nilpotency of semigroups are proved in detail. The lecture notes are an expanded version of lectures given by the author at the University of Edinburgh in 1980 and can be used as a text for a graduate course in functional analysis.

0 521 28598 4

For a list of books previously published in this series see page i